NEUTRAL CURRENT SHEETS IN PLASMAS
NEITRAL'NYE TOKOVYE SLOI V PLAZME
НЕЙТРАЛЬНЫЕ ТОКОВЫЕ СЛОИ В ПЛАЗМЕ

The Lebedev Physics Institute Series

Editors: Academicians N.G. Basov and D.V. Skobel'tsyn

P. N. Lebedev Physics Institute, Academy of Sciences of the USSR

Recent Volumes in this Series

In preparation

Proceedings (Trudy) of the P. N. Lebedev Physics Institute

Volume 74

NEUTRAL CURRENT SHEETS IN PLASMAS

Edited by
N. G. Basov
P. N. Lebedev Physics Institute
Academy of Sciences of the USSR
Moscow, USSR

Translated from Russian by
Dave Parsons

CONSULTANTS BUREAU
NEW YORK AND LONDON

Library of Congress Cataloging in Publication Data

Main entry under title:

Neutral current sheets in plasmas.

(Proceedings (Trudy) of the P. N. Lebedev Physics Institute; v. 74)
Translation of Neĭtral' nye tokovye sloi v plazme.
Includes bibliographical references and index.
1. Electromagnetism–Addresses, essays, lectures. 2. Plasma dynamics–Addresses, essays, lectures. I. Basov, Nikolaĭ Gennadievich, 1922- II. Series: Akademiia nauk SSSR. Fizicheskiĭ institut. Proceedings; v. 74.
QCl.A4114 vol. 74 [QC718.5.E45] 530'.08s [530.4'4]
ISBN 978-1-4615-8566-4 ISBN 978-1-4615-8564-0 (eBook) 76-17087
DOI 10.1007/978-1-4615-8564-0

The original Russian text was published by Nauka Press in Moscow in 1974 for the Academy of Sciences of the USSR as Volume 74 of the Proceedings of the P. N. Lebedev Physics Institute. This translation is published under an agreement with the Copyright Agency of the USSR (VAAP).

A Division of Plenum Publishing Corporation
227 West 17th Street, New York, N. Y. 10011

CONTENTS

NEUTRAL CURRENT SHEETS IN PLASMAS IN SPACE AND IN THE LABORATORY

S. I. Syrovatskii

The appearance of a neutral current sheet in a plasma with a strong magnetic field containing a null line is discussed. The properties of the sheet and its stability are discussed.

1. Introduction

The main reasons for studying neutral current sheets in plasmas are to determine whether the magnetic energy of the current in a current sheet can be converted into the kinetic energy of the directed motion of charged particles and, if this conversion is possible, to determine how efficient it is.* In other words, the purpose is to study the properties of a plasma accelerator in which energy is accumulated comparatively slowly near a current sheet and then liberated rapidly, much of it being converted into the kinetic energy of accelerated particles.

Work in this direction was spurred by a study of events in plasmas in space, parimarily in solar chromospheric flares, in which there is a rapid conversion of magnetic energy into the energy of accelerated particles in large volumes. In weak flares these particles are usually electrons, with energies ranging from a few keV to hundreds of keV; in strong chromospheric flares the energies of the electrons and nuclei reach hundreds of MeV or more.†

The typical temperatures, plasma densities, and magnetic fields in solar flares are similar to the corresponding properties of laboratory plasmas. For example, the temperatures $T = 10^4$-10^8 °K, densities $n = 10^{10}$-10^{14} cm^{-2}, and magnetic fields $H = 10^2$-10^3 Oe characteristic of flares are very similar to the typical properties of laboratory plasmas. The difference in linear dimensions, on the other hand, is more than seven orders of magnitude.

Because of this tremendous difference it is essentially impossible to satisfy the similarity conditions and thus exactly simulate in the laboratory the processes which occur at current sheets in space. Nevertheless, the study of analogous processes under laboratory conditions is by no means ruled out. According to the principle of limited modelling [4], it is not necessary to preserve the exact values of all the dimensionless parameters; it is suffi-

* This conversion has been called [1] the "dynamic dissipation of a magnetic field" in order to emphasize the distinction from "ordinary" dissipation, in which energy is converted into heat.

† According to recent data the energy of the nuclei can reach 100 GeV [2, 3].

cient to preserve only the order-of-magnitude relations among them. For example, if some dimensionless parameter R of a plasma in space satisfies the condition $R \gg 1$, then limited modelling requires that the corresponding parameter R' in the laboratory system also satisfy a condition $R' \gg 1$. The ratio R'/R is left arbitrary. One parameter whose value in space is usually impossible to reproduce in the laboratory is the magnetic Reynolds number

$$\mathrm{Re}_m = \frac{4\pi\sigma L V}{c^2},$$

in which the typical values of the plasma velocity V and conductivity σ in space and in the laboratory are comparable in order of magnitude, while the values of L differ by many orders of magnitude. Fortunately, it is possible to arrange the condition $\mathrm{Re}_m \gg 1$ in a laboratory plasma, although this inequality cannot be made as strong as in space. This circumstance is the basis for the emergence of experiments designed to study flare processes at current sheets. Indirect evidence that such processes operate in the laboratory comes from several observations of accelerated particles with nonthermal energy spectra in discharges of the Θ-pinch type [5]. A characteristic feature of such discharges is a rapidly evolving current sheet. We do not rule out the possibility that the same processes occur in the later stages of the discharge in a cylindrical Z pinch, in which the plasma filament becomes unstable, and large deviations from axial symmetry appear.

Study of the possibility of particle acceleration at a current sheet involves two main steps: producing a stable current sheet and causing its rupture. We will consider these two steps in turn.

2. Production of a Quasisteady Neutral Sheet

The processes which occur near neutral points and neutral sheets in plasmas have been the subject of a considerable number of theoretical papers, beginning with the papers by Dungey and Sweet (see [6, 7] and the literature cited there). The appearance of a neutral current sheet has been studied in detail, with physically justified boundary conditions, for plasma motion near a magnetic null line [1, 8-12]. Early in this work [1, 8] it was shown that there is a broad range of plasma motions which can cause a cylindrical wave near a null line and a progressive increase in the current near the null line.

The results of [1, 8] are valid only in the linear, small-perturbation approximation, and the extrapolation to the nonlinear region used in those papers is not valid, although it does indicate correctly the directions in which various properties change. As was shown in [9-12], the current in the nonlinear stage takes the form of a thin sheet which separates regions in which the magnetic fields are equal in magnitude but opposite in direction; i.e., a neutral current sheet arises. These papers suggested a method for producing a neutral current sheet in the laboratory: A uniform electric field E is to be applied along the axis of a magnetic quadrupole, which has the lowest-order null line [13, 14]. Far from the null line this field causes a plasma drift corresponding to the simplest case of a converging cylindrical wave.

The quadrupole method is of course not the only possible method, and it may not be the most efficient in terms of the ultimate goal of particle acceleration. For example, extended current sheets can be produced comparatively easily in the quasi-one-dimensional (the only dependence is on r) geometry of Θ pinches [15]. An interesting experiment with a double inverse pinch has apparently also produced a current sheet [16, 17]. An advantage of the quadrupole method is that all the characteristic parameters — the external magnetic field, the electric field producing the sheet, and the plasma density — can be adjusted independently and over broad ranges. This capability makes the quadrupole method particularly valuable for the early stage of the study.

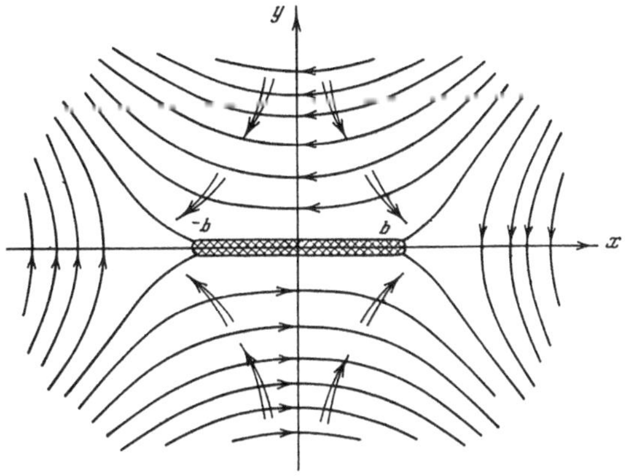

Fig. 1. Current sheet in an external hyperbolic magnetic field. The double arrows show the plasma flow in the presence of an electric field E.

To review the theoretical results and estimates for the simplest case, that of a neutral current sheet arising from a first-order null line, we use the Gaussian unit system. We denote by h_0 the gradient of the external magnetic field along the null line, by H_s the magnetic field on both sides of the sheet, by 2b the sheet width (Fig. 1), and by $2a$ its thickness. Near such a sheet in the limit $a \to 0$ the magnetic field can be expressed in terms of the complex potential [10]

$$F(z,\ t) = \frac{h_0}{2} z \sqrt{z^2 - b^2} - \frac{2I}{c} \ln \frac{z + \sqrt{z^2 - b^2}}{b} + A_0(t) \tag{1}$$

[where z = x + iy, I is the total current in the sheet, and A(t) is the magnetic flux dissipated in the sheet by time t] in the following manner:

$$H_x - iH_y = i\frac{dF}{dz} = h_0 \sqrt{b^2 - z^2}. \tag{2}$$

Here and below we assume a quasisteady regime in which there are no return currents in the sheet. In this case we can write [10]

$$I = \frac{ch_0 b^2}{4}. \tag{3}$$

Green [18] found the magnetic field for the particular case of two parallel currents separated by a transverse neutral sheet. To determine the sheet width b he used the condition that the external currents are at equilibrium (i.e., that there is no resultant Maxwell stress in the symmetry plane containing the current sheet). This is not a necessary condition, since the sources of the external magnetic field can be maintained by external forces. Accordingly, the sheet width can, generally speaking, be arbitrary, governed by the conditions under which the sheet is formed.

It follows from Eq. (2) that near the sheet (x < b, y $\simeq$ 0) we have

$$H_y \simeq 0, \qquad H_x \simeq h_0 \sqrt{b^2 - x^2}$$

and that the magnetic field is quasiuniform over a large part of the region $|z| < b$:

$$H_x \simeq H_s \equiv h_0 b. \tag{4}$$

We will use this relationship between the field near the sheet and the sheet width b in estimates below. We note that the profile of the surface current in the sheet, $\mathcal{J}(x)$, is like the H(x) profile:

$$\mathcal{J}(x) = \frac{ch_0}{2\pi} \sqrt{b^2 - x^2}. \tag{5}$$

Under conditions such that we can neglect the resistance of the sheet (the case of an ideal conductivity), the width of the sheet must, according to [10], continuously increase as time elapses, approximately in proportion to $\sqrt{t}$ in a constant external field E. This result follows from the expression for b in terms of the field potential at the null line (see Eq. (52) in [101]):

$$b^2 \simeq \frac{4\beta(t)}{h_0\,[1+\ln(h_0 l^2/4\beta(t))]}, \qquad \beta(t) = c\int_0^t E\,dt. \tag{6}$$

Here l is the scale dimension of the contour closing the current in the current sheet.

Actually, however, the sheet resistance becomes significant very rapidly, and the sheet width approaches the steady-state value b = const ($\beta(t) \to A(t)$ + const in Eq. (52) in [10], where A(t) is the magnetic flux dissipated in the sheet).

Equations (1)-(6) were derived for the case of a high plasma conductivity [see condition (9) below] and for the inequalities

$$H^2/4\pi \ll nmc^2, \tag{7}$$

$$H^2/4\pi \gg nkT, \tag{8}$$

where m is the electron mass, n is the density, k is the Boltzmann constant, and T is the temperature. The first of these inequalities ensures the applicability of the MHD approximation and is equivalent to the condition $(\omega_H/\omega_0)^2 \ll 1$, where ω_0 is the electron plasma frequency and ω_H is the electron gyrofrequency. The second inequality corresponds to the strong-field approximation [10], in which the fields can be assumed a potential field everywhere except at the thin neutral sheet.

It is not difficult to obtain an estimate for the steady-state sheet width b or, more precisely, for the ratio of the sheet width to the sheet thickness, b/a. For this purpose we use Ohm's law $j = \sigma E$, which must hold in the steady state, and we use Eq. (4) to write the current density as $j = (c/4\pi)/|\mathrm{rot}\,\mathbf{H}| \approx cH_s/4\pi a$. We then find

$$b/a - 4\pi j/ch_0 = 4\pi\sigma E/ch_0 \tag{9}$$

Estimate (9) also follows from the condition for a steady state, in which the plasma drift velocity toward the sheet,

$$V_d = cE/H_s, \tag{10}$$

is equal to the rate at which the field penetrates into the plasma,

$$V_\sigma = c^2/4\pi\sigma a. \tag{11}$$

To estimate the thickness of the sheet, a, we use the momentum conservation equation, which reduces in the steady state to the pressure equation

$$\frac{H_s^2}{8\pi} = nk(T_e + T_i), \tag{12}$$

where n is the plasma density at the center of the sheet, and T_e and T_i are the electron and ion temperatures in the direction across the sheet ($kT = mv_\perp^2/2$). Using Eq. (12) along with the equation $\mathrm{rot}\,\mathbf{H} = (4\pi/c)/j$, i.e.,

$$\frac{H_s}{a} \simeq \frac{4\pi}{c} nev_s, \tag{13}$$

where v_s is the average electron current velocity in the sheet (we neglect the ion current velocity), we find the following equation for the sheet thickness:

$$a = r_{H_e} \frac{v_{T_e}}{v_s}\left(1 + \frac{T_i}{T_e}\right), \tag{14}$$

where $v_{T_e} = (2kT_e/m)^{1/2}$, and where $r_{He} = mcv_{T_e}/eH_s$ is the electron gyroradius in field H_s. For given values of the field H_s, the electron temperature T_e, and the ion temperature T_i, the sheet thickness a depends on the directed velocity v_s. The latter, according to (13), can be expressed in terms of the total number of particles per unit surface area of the sheet,

$$N = an, \tag{15}$$

by

$$v_c = \frac{cH_s}{4\pi eN}. \tag{16}$$

Then for given values of H_s, T_e, and T_i, the sheet thickness decreases with a decrease in the total number of particles in the sheet; this result will play an important role in the discussion of the stability of the sheet.

3. The Neutral Sheet or Petschek Flow?

The concept of the appearance of a neutral sheet is not universally accepted. The Petschek model [21] has been used in many papers in discussions of the processes occurring near a null line (see, e.g., [16, 17, 19, 20]). According to this model, four slow-mode MHD shock waves depart from a null line. In these shock waves, the slow plasma flow toward the null line in one pair of opposite sectors is transformed into a rapid flow away from the null line in the other pair of sectors. The immediate vicinity of the null line is the so-called diffusion region, in which the magnetic lines of force of the two oppositely directed plasma flows are reconnected.

The Petschek diffusion region differs from a neutral sheet in two regards: First, the current density at the center of this region is minimal, while that at the center of the neutral sheet is maximal [see (5)]. Second, as the conductivity increases, the width of the diffusion region should decrease (Eq. (20) of [21]), while the width of a quasisteady neutral sheet increases [see (9)].

The Petschek model has been subjected to some criticism [22-24]; to meet this criticism it is necessary to (at least) complicate the model considerably by introducing an additional set of waves [25]. If this measure is taken, it is necessary to determine whether this complicated pattern can be reconciled with plausible external boundary conditions; furthermore, there is the problem of the very establishment of this type of flow – a problem which has received no attention at all.

In contrast, the problem of the neutral sheet near a null line is formulated from the very beginning as an unsteady boundary-value problem. Significantly, the solution does not generally reach a steady state. When the finite conductivity is taken into account, the magnetic field outside the sheet ultimately assumes the steady-state configuration in (2) with a constant value of b, but the hydrodynamic flow remains fundamentally unsteady, as can be seen in the progressive decrease of the density near the sheet (see [12] and the discussion below).

Yet another distinction between the neutral sheet and Petschek flow lies in the nature of the boundary conditions. In the neutral-sheet case the field far from the sheet is assumed to be approximately a potential field corresponding to the null point, and this field is described by a gradient h_0. In the Petschek model a uniform magnetic field H_0 is specified in the incoming flow, and the distortions of this field caused by the shock waves in the diffusion region are assumed small.

This assumption is apparently the most vulnerable part of the Petschek model: Study of the exact steady-state flows of an incompressible fluid shows [26-29] that the steady-state solution should include at least one system of waves in addition to the slow-mode Petschek waves. A question which remains unanswered is whether the solution can be generalized to the case of a compressible fluid, since these additional waves correspond to nonevolving rarefaction shocks. It is extremely probable that no steady-state solution exists at all for general boundary conditions, as can be seen from the example of flow near a steady-state neutral sheet (which we examine below).

Furthermore, numerical calculations for hydrodynamic plasma flow near a null line offer no support for the Petschek model [11, 30, 31]. These calculations demonstrate the appearance and development of a neutral sheet, but they give no indication of the appearance of a system of slow-mode Petschek shock waves.

Finally, the appearance of a neutral sheet at a preexisting null line has been observed experimentally [13, 14, 32, 33]. In particular, a special study in [33] revealed no transverse component of the magnetic field (of course, within the experimental error), while such a component should exist in the Petschek diffusion region.

On the other hand, some investigators assert [16, 17, 34] that a system of slow-mode Petschek shock waves has been observed in their experiments. However, these investigators themselves emphasize an important discrepancy: The current has a maximum at the null line instead of the minimum which should be there according to the Petschek model. Actually, if we use Eq. (9) and the experimental parameters of [16] ($j = 5.8$ kA/cm^2, $I = 6 \cdot 10^4$ A, and $h_0 = 0.4\ I/a^2 \simeq 10^3$ G/cm at time $t = 7$ msec), we find that b/a, the ratio of the sheet width and thickness, is 7.6, in good agreement with the size of the current region bounded by the contour with the current density $j = 2$ kA/cm^2 (see Fig. 3 in [16]). Accordingly, it would be more natural to assume that it was actually a neutral sheet, not very large, which was observed in the experiments of [16, 17], but that the situation was distorted by the high density of the displaced plasma.

Accordingly, there is enough evidence for us to conclude that, with natural assumptions regarding the boundary conditions, the question can be resolved in favor of the neutral sheet. At any rate, this conclusion has been demonstrated quite rigorously for a homogeneous plasma in a strong hyperbolic magnetic field near a null line. It is precisely this situation which is most frequently encountered in plasmas in space, in which the neutral sheets arise when inequality (8) holds. We emphasize that, because of this inequality, shock waves, if they appear in the flow, must be weak and can have little influence on the flow.

On the other hand, under other conditions (with a high-density plasma, particular boundary conditions, or certain processes within the thickness, i.e., "within" the neutral sheet) our arguments lose their validity, and the Petschek model [21] or a generalization of it may turn out to be applicable [25-29].

4. Decrease in the Plasma Density near a Neutral Sheet

One of the important properties of plasma flow near a neutral current sheet is a monotonic decrease in the density of the incoming plasma. This feature was pointed out in [1], but the conclusions of that paper referred to small perturbations only. Numerical calculations show that the density also decreases in a region of nonlinear flow near a sheet; for a steady-state sheet, the density falls off monotonically at each point near the sheet surface [12].

For a steady-state sheet [b = const, A(t) = −cEt, where E is the external electric field] we can easily find the asymptotic behavior of the density at large t. For this purpose we consider the two-dimensional plasma flow near a neutral sheet in the approximation of a strong magnetic field; this flow is described by [10]

$$\Delta A = 0, \qquad \frac{dA}{dt} = 0, \qquad \frac{d\mathbf{v}}{dt} \times \nabla A = 0, \qquad \frac{d\rho}{dt} = -\rho \operatorname{div} \mathbf{v}. \tag{17}$$

The solution of the first equation is the real part of the potential in (1): $A \equiv A(x, y, t) = \operatorname{Re} F(z, t)$. Here the neutral sheet is assumed infinitesimally thin and is treated as a cut on the complex plane.

We consider the flow in the region $x \ll y$, $b \ll y$, i.e., near the y axis and far from the sheet. Retaining the leading terms in the expansions of potential (1) and of its derivatives in terms of the small quantities $(x/y)^2$ and $(b/y)^2$, we find

$$A(x, y, t) = \frac{h_0}{2}\left[x^2 - y^2 - \frac{b^2}{2}\left(1 + 2\ln\frac{2y}{b}\right)\right] + A(t) \simeq -\frac{h_0}{2}y^2 + A(t),$$
$$\frac{\partial A}{\partial x} = h_0 x, \quad \frac{\partial A}{\partial y} = -h_0 y. \tag{18}$$

As a result, the second and third equations in (17) become

$$y^2 = y_0^2 - \frac{2cE}{h_0}t, \qquad \ddot{x} + \frac{\ddot{y}}{y}x = 0. \tag{19}$$

Here x and y are treated as Lagrange coordinates, and x_0 and y_0 are their initial x values (at t = 0). Transforming to the variable $\eta = y^2 = y_0^2 - \lambda t$, $d/dt = -\lambda(d/d\eta)$, where $\lambda = 2cE/h_0$, we find an equation for x(t):

$$\frac{d^2x}{d\eta^2} - \frac{x}{4\eta^2} = 0. \tag{20}$$

This equation can be solved without difficulty, and at large values of t, with $y_0^2 \simeq \lambda t \gg y^2$, the solution is

$$x \simeq x_0\left(\frac{y_0}{y}\right)^{\sqrt{2}-1}. \tag{21}$$

The last equation in (17) is conveniently written in the Lagrange form $\rho_0 = \rho[D(x, y)/D(x_0, y_0)]$, from which, using (19) and (21), we find

$$\frac{\rho}{\rho_0} = \left(\frac{y}{y_0}\right)^{\sqrt{2}} \simeq \left(\frac{y^2}{\lambda t}\right)^{\sqrt{2}/2}. \tag{22}$$

This equation describes the density behavior at a point far from the sheet ($y^2 \gg b^2$). In the subsequent plasma motion toward the neutral sheet (in the region $y \lessapprox b$), the density (at any rate) does not increase. Accordingly, to find an upper estimate for the density near the current sheet we can use

$$\rho \leqslant \rho_0\left(\frac{h_0 b^2}{cEt}\right)^{\sqrt{2}/2}. \tag{23}$$

It follows that for sufficiently prolonged motion in a constant electric field the plasma density near the neutral sheet decreases without bound.

5. Stability of a Neutral Sheet

The magnetic energy associated with a neutral sheet is proportional to the square of the total current in the sheet, I. Accordingly, as follows from (3), it is advantageous to have a sheet of maximum width. However, even with a comparatively small width-to-thickness ratio,

$$\frac{b}{a}=\frac{4\pi\sigma E}{ch_0}>2\pi\,, \tag{24}$$

the electromagnetic tearing-mode instability is driven in the sheet and leads to a reconnection of the magnetic lines of force through the sheet [35-39].

This instability is driven in both collisional plasmas (in the hydrodynamic approximation) and collisionless plasmas. It is therefore unclear how an extended current sheet can develop and exist in a quasisteady state, whether it be in the laboratory [14-16], in the magnetospheric tail [40], or in a solar flare [30, 41-44]. An answer to this question would be of fundamental importance for the entire problem, since it would furnish an explanation for the conditions under which the energy is stored in the sheet and the conditions under which this energy is rapidly liberated. Below we attempt a qualitative analysis of this question and point out approaches to its solution.

We consider a neutral sheet whose thickness is large in comparison with the mean free path of the particles and for which we can use the hydrodynamic approximation. These conditions are usually satisfied in the laboratory and in chromospheric flares, which occur in an atmosphere with a relatively high initial plasma density.

The instability of a sheet has been demonstrated in the linear hydrodynamic approximation under the assumption that there is no plasma outside the sheet. These two restrictions are apparently responsible for the discrepancy between theory and observation: The tearing-mode instability leads to a breakup of the neutral sheet into current filaments and, correspondingly, to the appearance of magnetic null points within the sheet. As long as this process is occurring within the sheet (the perturbations are small in the linear approximation), no contradictions arise; the theory is applicable, and it shows that the sheet structure is generally unsteady. The question of importance to us, on the other hand, is whether an instability of this type can lead to the breakup of the sheet as a whole.

Such a breakup is clearly possible only if the sheet is in vacuum, i.e., if condition (7) does not hold outside the sheet. Otherwise, at the points at which the sheet breaks up, in the low-pressure plasma [condition (8) is assumed to hold outside the sheet], singluar null points with $H = 0$ but $E \neq 0$ would appear outside the sheet [10, 12]. Here the field E would consist of the uniform external field, which maintains the current in the sheet, and a transient electric field which arises during the breakup of the sheet [39, 45]. As was shown in [10], singular null points cannot exist in a plasma: If such a point appeared, a current sheet would develop at this point at a velocity on the order of the Alfvén velocity (see [9]). We emphasize that this latter process is highly nonlinear, so that we can say that, if conditions (7) and (8) hold, there is a nonlinear stabilization of the hydrodynamic tearing-mode instability of a neutral sheet.

This result is justified to the same extent as the appearance of a neutral sheet from a null line. On the one hand, it explains the stability of a neutral sheet against breakup if the sheet is surrounded by a plasma "coat." Then within a sheet satisfying condition (24) there will be some noise due to the tearing-mode instability which does not lead to a breakup of the sheet as a whole. In particular, this means that the internal structure of the sheet itself can be very complicated and can contain both null points and a transverse component of the magnetic field, under the sole condition that this component is small in comparison with the longitudinal component [see (4)].

On the other hand, it becomes clear that there can be a rapid breakup or change in the configuration of the sheet under certain conditions. Let us examine these conditions.

First, it is clear from the preceding arguments that the sheet breaks up if, as a result of a sufficiently prolonged effect of the field E, the plasma density in the incoming flow, given by (22), falls below limit (7), while condition (24) remains satisfied. In this case, nonlinear hydrodynamic stabilization is ruled out, and the sheet as a whole must break up. This is a very interesting situation, since it would permit us to obtain an evacuated region with a strong electric field within a plasma [1, 39, 45]. However, the occurrence of such a breakup, although extremely probable, has not yet been demonstrated either theoretically or experimentally.

Second, the sheet configuration must change rapidly, according to (9), upon a rapid decrease in the conductivity resulting from the appearance of plasma turbulence. This turbulence arises if, as a result of the escape of plasma along the sheet and the resulting decrease in N, the electron current velocity in (16) reaches the threshold for the launching of plasma waves of one type or other. Ion-acoustic turbulence is most probable; for a plasma with $T_i \gtrsim T_e$ it sets in at $v_s \geq v_{T_e}$. The role of this type of turbulence for flow near a neutral sheet was discussed in [19, 20, 44, 46, 47]. It has apparently been observed experimentally [16, 32, 33, 48-50]. The rapid decrease in the conductivity in certain parts of the sheet is actually equivalent to the breakup of the sheet. At any rate, under quasisteady external conditions, the sheet must undergo a change in shape in accordance with the new value of σ [see (9)]. Just such a change in shape has apparently been observed experimentally [16, 17, 48], since "vacuum" breakup of the sheet in these experiments was ruled out by the high density in the incoming plasma flow ($n \simeq 4 \cdot 10^{13}$ cm^{-3} [16]). With regard to the experiments of [49], we note that the turbulent nature of the conductivity prevented the development of a sheet from the very beginning, in accordance with the parameters of this experiment and in accordance with (9) (see [33] for more details).

We note that the sheet can retain a large width-to-thickness ratio even after the change in configuration, if the new "turbulent" value of σ still satisfies condition (24). A situation of this type is difficult to arrange in the laboratory, but we see nothing to prevent it in situations in space because of the large distances and small values of h_0 involved (in the laboratory, a lower limit is set on h_0 by the smallness of the ion gyroradius). In this case a sheet with developed ion-acoustic turbulence exists until the external field E decays or, if this field is maintained for a sufficiently long time, until the breakup occurs as a result of the continuing decrease in the plasma density near the sheet. Accordingly, in contradiction of the conclusion reached in [47], ion-acoustic turbulence is not an obstacle to sheet breakup or to the concomitant particle acceleration if these conditions are satisfied.

We have been discussing the behavior of a neutral sheet in terms of instabilities, which is a convenient approach in this case since it permits us to divide the very complicated situation into two stages: the establishment of the sheet and its breakup. However, we achieve this simplification only at the expense of losing certain important features of the flow, since in this analysis of the stability we are assuming homogeneity along the sheet and we are making arbitrary assumptions regarding the nature of the initial perturbations. Actually, we are dealing with a common flow in which an inhomogeneous neutral sheet is formed; at certain places in this sheet, at a certain stage of the flow, a process begins which is interpreted as breakup. In this sense we may speak, as in [1], not in terms of an instability of the sheet, but in terms of the appearance of regions with a very low density and a high current density (a high field gradient) during flow near a null line. It is these regions which are the breakup points.

In conclusion, we should say a few words about a collisionless neutral sheet. In this case the situation is less clear, both theoretically and experimentally. Biskamp et al. [51] have shown that there is pronounced stabilization of the tearing-mode instability in the nonlinear (quasilinear) stage. This result holds for a thick sheet ($r_{H_e} \ll a$ and, probably, $r_{H_i} \ll a$ [39]) and can explain the stability against the breakup of a collisionless neutral sheet, e.g., in the magnetospheric tail. However, we are left with the question of the reason for the rapid change in the structure of this sheet, e.g., during magnetospheric substorms. It is possible that the external "plasma coat" and the influx of plasma into the sheet under the influence of the electric field are playing a role here. Nor do we rule out the possibility that in a sheet with $a \lessgtr r_{H_i}$ the polarization of the plasma due to its drift under the influence of the electric field is important. Accordingly, we can speak of quasilinear stabilization [51] only for a thick sheet, with $a \gg r_{H_i}$ (see also [39]). Finally, the influence of the ion-acoustic instability in a thin sheet has not been resolved. It may turn out that the situation is no different from the hydrodynamic situation in this case. It is difficult to experimentally simulate a collisionless sheet because at a low density it is difficult to achieve the high plasma conductivity required for the appearance of a sheet.

It is clear from this discussion that, depending on the initial and boundary conditions, there are at least two different modes in the behavior of a developed current sheet. One mode is characterized by the onset of ion-acoustic turbulence, plasma heating, and, perhaps, the appearance of some runaway particles. In the second mode the sheet breaks up, and there is an effective acceleration of particles by the pulsed electric field.

The variety of conditions under which each of these modes can occur and the various combinations of these modes in a single event are responsible for the extremely complex behavior of neutral sheets. This complexity is particularly familiar in the case of solar flares [52].

Literature Cited

1. S. I. Syrovatskii, Astron. Zh., 43:340 (1966).
2. S. M. Schindler and P. D. Kearney, Nature, 237:503 (1972).
3. S. M. Schindler and P. D. Kearney, Nature Phys. Sci., 242:56 (1973).
4. I. M. Podgornyi, É. M. Dubinin, and G. G. Managadze, Preprint D-15, Institute of Cosmic Studies, Academy of Sciences of USSR, 1970.
5. L. A. Artsimovich, Controlled Thermonuclear Reactions, Gordon and Breach, New York (1968).
6. J. W. Dungey, Cosmic Electrodynamics, Cambridge Univ. Press (1958).
7. P. A. Sweet, in: Annual Review of Astronomy and Astrophysics, Vol. 7 (1969), p. 149.
8. S. I. Syrovatskii, Zh. Éksp. Teor. Fiz., 50:1133 (1966).
9. V. S. Imshennik and S. I. Syrovatskii, Zh. Éksp. Teor. Fiz., 52:990 (1967).
10. S. I. Syrovatskii, Zh. Éksp. Teor. Fiz., 60:1727 (1971).
11. N. I. Gerlakh and S. I. Syrovatskii, Tr. FIAN, 74:73 (1974).
12. B. V. Somov and S. I. Syrovatskii, Tr. FIAN, 74:14 (1974).
13. S. I. Syrovatskii, A. G. Frank, and A. É. Khodzhaev, ZhÉTF Pis. Red., 15:138 (1972).
14. N. Ohyabu and N. Kawashima, J. Phys. Soc. Jpn, 33:496 (1972).
15. M. Alidieres, R. Aymar, P. Jourdan, F. Koechlin, and A. Samain, Plasma Phys., 10:841 (1968).
16. A. Bratenahl and C. M. Yeates, Phys. Fluids, 13:2696 (1970).
17. P. J. Baum and A. Bratenahl, Preprint, June, 1973.
18. R. M. Green, IAU Symp. No. 22, 1963. Stellar and Solar Magnetic Fields (ed. R. Lust), North-Holland, Amsterdam (1965), p. 398.
19. M. Friedman and S. M. Hamberger, Astrophys. J. 152:667 (1968).
20. M. Friedman, Phys. Rev., 182:1408 (1969).

21. H. E. Petschek, Proceedings of the AAS-NASA Symposium on the Physics of Solar Flares, Washington, 1964, p. 423.
22. R. M. Green and P. A. Sweet, Astrophys. J. 147:1153 (1967).
23. C. P. Sonett, Comments on Astrophys. Space Phys., 1:87 (1969).
24. E. R. Priest, Monthly Not. Roy. Ast. Soc., 159:389 (1972).
25. H. E. Petschek and R. M. Thorne, Astrophys. J., 147:1157 (1967).
26. B. U. O. Sonnerup, J. Plasma Phys., Part 1, 4:161 (1970).
27. T. Yeh and W. I. Axford, J. Plasma Phys. 4:207 (1970).
28. E. R. Priest, Astrophys. J., 181:227 (1973).
29. T. Yeh and M. Dryer, Astrophys. J., 182:301 (1973).
30. J. C. Stevenson, J. Plasma Phys., 7:293 (1972).
31. S. I. Syrovatskii, in: Solar-Terrestrial Physics, 1970 (ed. E. R. Dyer), Reidel, Dordrecht (1972), p. 119.
32. S. I. Syrovatskii, A. G. Frank, and A. Z. Khodzhaev, in: Proceedings of 5th European Conference on Controlled Fusion and Plasma Physics. Grenoble, France, 1972, Vol. 1, p. 150.
33. A. G. Frank, Tr. FIAN, 74:108 (1974).
34. A. Bratenahl, "Experimental studies of the reconnection process," Invited Paper, AGU Symposium on the Geomagnetic Tail: Dynamics and Topology, San Francisco, December 8, 1971.
35. H. P. Furth, I. K. Killeen, and M. N. Rosenbluth, Phys. Fluids, 6:459 (1963).
36. G. Laval, R. Pellat, and M. Vuillemin, Proceedings of the Conference on Plasma Physics and Controlled Nuclear Fusion, Culham, 1965, Vienna (1966), Vol. II, p. 259.
37. V. K. Neil, Phys. Fluids, 5:14 (1962).
38. M. A. Gross and G. van Hoven, Phys. Rev. A4:2347 (1972).
39. S. V. Bulanov and S. I. Syrovatskii, Tr. FIAN, 74:88 (1974).
40. B. Coppi, G. Laval, and R. Pellat, Phys. Rev. Letters, 16:1207 (1966).
41. R. K. Jaggi, J. Geophys. Res., 68:4429 (1963).
42. P. A. Sturrock, Nature, 211:695 (1966).
43. P. A. Sturrock, Solar Phys., 23:438 (1972).
44. S. I. Syrovatskii, in: Solar Flares and Space Research, North-Holland, Amsterdam (1969), p. 346.
45. S. I. Syrovatskii, Izv. Akad. Nauk SSSR, Ser. Fiz., 31:1303 (1967).
46. M. Friedman and S. M. Hamberger, Solar Phys., 8:104 (1969).
47. D. F. Smith and E. R. Priest, Astrophys. J., 176:487 (1972).
48. P. J. Baum and A. Bratenahl, "Spectrum of turbulence at a magnetic neutral point," preprint, October 1973.
49. S. I. Syrovatskii, A. G. Frank, and A. E. Khodzhaev, Zh. Tekh. Fiz., 43:912 (1973).
50. M. V. Babykin, A. I. Zhuzhanashvili, and S. S. Sobolev, Zh. Éksp. Teor. Fiz., 60:345 (1971).
51. D. Biskamp, R. Z. Sagdeev, and K. Schindler, Cosmic Electrodynamics, 1:297 (1970).
52. S. I. Syrovatskii, Comments Astrophys. Space Phys., 4:65 (1972).

HYDRODYNAMIC PLASMA FLOW IN A STRONG MAGNETIC FIELD

B. V. Somov and S. I. Syrovatskii

Several plasma-dynamics problems, pertaining primarily to space physics, are analyzed in the approximation of a strong magnetic field. These problems can be grouped into two classes: continuous flow without magnetic null points and flow leading to the appearance of current sheets in the presence of singular null points. Flow in the field of a magnetic dipole which is increasing with time – a problem of the first class – is analyzed. There is a discussion of the application of the results to topics in solar physics – the formation of prominences, condensations, spicules, etc. A problem of the second class which is discussed is that of the appearance of a current sheet in the field of a plane magnetic dipole. There is a discussion of the application of these results to the problem of the appearance of the geomagnetic tail and coronal rays. The particle trajectories and the velocity and density profiles near a developed current sheet are found by numerical integration. An important result is the proof that the plasma density near a current sheet decays over time.

INTRODUCTION

Some approximation or other is usually used to solve the nonlinear problems of magnetohydrodynamics [1-4]. A particularly common approximation is the approximation of a weak magnetic field, in which the magnetic effects are treated as small corrections to the hydrodynamic effects. Many problems of cosmic electrodynamics and plasma physics have been solved in this approximation.

Among the simplest of these problems are those dealing with the influence of a weak magnetic field on a hydrostatic equilibrium; examples are the influence of weak poloidal [5, 6] and toroidal [7, 8] magnetic fields on the equilibrium of a self-gravitating plasma sphere (a star).

Essentially analogous problems are those which might be called "kinematic" (following [9]) since they deal with only one aspect of the interaction between a plasma and a magnetic field – the effect of a given plasma flow on the magnetic field (the inverse effect is assumed negligible). Such problems reduce to finding the magnetic field profile resulting from the given velocity field. Examples are the problems of the intensification and maintenance of a magnetic field by steady-state plasma flows (dynamos) [10, 11] and of the turbulent intensification of a magnetic field [12, 13]. The simplest examples are problems on the intensification of a magnetic field by differential plasma rotation [14, 15].

An approximation which has received less study, although it better reflects the peculiar features of magnetohydrodynamics, is the opposite approximation – the approximation of a strong magnetic field, proposed and used in* [16-19]. This approximation, in contrast to that of a weak field, is applicable when the magnetic force is predominant over all other forces: the gravitational force, the gas-pressure gradient, etc., including the inertial force.

We start from the system of MHD equations for an ideal medium [1, 2]:

$$\frac{\partial \mathbf{v}}{\partial t} + (\mathbf{v}\nabla)\,\mathbf{v} = -\frac{\nabla p}{\rho} - \frac{1}{4\pi\rho}[\mathbf{H}\,\mathrm{rot}\,\mathbf{H}], \tag{1}$$

$$\frac{\partial \mathbf{H}}{\partial t} = \mathrm{rot}\,[\mathbf{v}\mathbf{H}], \tag{2}$$

$$\frac{\partial \rho}{\partial t} + \mathrm{div}\,\rho\mathbf{v} = 0, \tag{3}$$

$$\frac{\partial s}{\partial t} + \mathbf{v}\nabla s = 0, \tag{4}$$

$$\mathrm{div}\,\mathbf{H} = 0, \tag{5}$$

$$p = p(\rho,\ s). \tag{6}$$

Here and below $\mathbf{v}$ and ρ are the velocity and density of the plasma, treated as a continuum, p and s are the pressure and the entropy, and H is the magnetic field intensity. Otherwise the notation is standard.

We denote by L, T, V, ρ_0, p_0, S, H_0 the scale values of the length, time, velocity, density, pressure, entropy, and magnetic field intensity, respectively. Transforming to the dimensionless variables $r^* = r/L$, $t^* = t/T$, ..., $H^* = H/H_0$ in Eqs. (1)-(6), we find the following system of dimensionless equations (below we omit the asterisk) [16-18]:

$$\frac{\varepsilon^2}{\delta}\frac{\partial \mathbf{v}}{\partial t} + \varepsilon^2(\mathbf{v}\nabla)\,\mathbf{v} = -\gamma^2\frac{\nabla p}{\rho} - \frac{1}{\rho}[\mathbf{H}\,\mathrm{rot}\,\mathbf{H}], \tag{7}$$

$$\frac{\partial \mathbf{H}}{\partial t} = \delta\,\mathrm{rot}\,[\mathbf{v}\mathbf{H}], \tag{8}$$

$$\frac{\partial \rho}{\partial t} + \delta\,\mathrm{div}\,\rho\mathbf{v} = 0, \tag{9}$$

$$\frac{\partial s}{\partial t} + \delta\mathbf{v}\nabla s = 0, \tag{10}$$

$$\mathrm{div}\,\mathbf{H} = 0, \tag{11}$$

$$p = p(\rho,\ s). \tag{12}$$

Here

$$\delta = VT/L, \qquad \varepsilon = V/V_A, \qquad \gamma^2 = p_0/\rho_0 V_A^2 \tag{13}$$

are characteristic dimensionless parameters of the problem, and $V_A = H_0/(4\pi\rho_0)^{1/2}$ is the scale value of the Alfvén velocity.

---

*See also the first paper in this collection and the literature cited there.

We call the magnetic field "strong" and we call the approximation corresponding to the given conditions *the approximation of a strong magnetic field* [16-19] if

$$\gamma^2 \ll 1, \qquad \varepsilon^2 \ll 1, \tag{14}$$

i.e., if the magnetic energy density is much larger than both the thermal and kinetic energy densities of the plasma.

In other words, the approximation of a strong magnetic field corresponds to the situation in which the magnetic force predominates over all other forces. It follows from this assumption [and formally from the equation of motion (7)] that a very strong magnetic field ($\gamma^2 \to 0$, $\varepsilon^2 \to 0$) must balance itself. We thus conclude that *in zeroth order in the small parameters in (14) the magnetic field is force-free*, i.e., satisfies the equation

$$[\mathbf{H} \operatorname{rot} \mathbf{H}] = 0. \tag{15}$$

If, on the other hand, there are no currents in zeroth order in γ^2 and ε^2, for example, because of the symmetry of the problem, a strong magnetic field is simply a potential field. A magnetic field can of course be force-free for another reason—because of an equilibrium among the nonmagnetic forces.

In first order in the small parameters in (14) the gas-pressure gradient obviously leads to a situation such that the magnetic field always differs from the force-free (potential) field: The magnetic tension must balance the gradient of the gas pressure as well as the magnetic pressure,

$$(\mathbf{H}\nabla)\mathbf{H} = \nabla\left(\frac{H^2}{2} + \gamma^2 p\right). \tag{16}$$

This effect is proportional to γ^2.

The inertial force, like the gas-pressure gradient, causes a deviation from a force-free (potential) magnetic field:

$$\varepsilon^2\left[\frac{1}{\delta}\frac{\partial \mathbf{v}}{\partial t} + (\mathbf{v}\nabla)\mathbf{v}\right] = -\frac{1}{\rho}[\mathbf{H} \operatorname{rot} \mathbf{H}]. \tag{17}$$

The deviation of the magnetic field from a force-free (potential) field is proportional to ε^2.

The first of the cases mentioned above evidently corresponds to the following relations between the parameters in (14):

$$\varepsilon^2 \ll \gamma^2 \ll 1. \tag{18}$$

In the second case, in which the gas-pressure gradient is neglected in first order, we have the opposite relation between the small parameters:

$$\gamma^2 \ll \varepsilon^2 \ll 1. \tag{19}$$

Since we will be interested in the hydrodynamic plasma flow in a strong magnetic field, we assume that the inequalities (19) hold, and we use Eq. (17) as the MHD equation of motion. It is natural to call the approximation corresponding to conditions (19) the *approximation of a strong magnetic field and a cold plasma* [16-19].

As we see from the equation of motion (17), the parameter $\delta = VT/L$ is a measure of the relative role of the local $(\partial/\partial t)$ and moving $(\mathbf{v}\nabla)$ terms in the substantial derivative d/dt.

If $\delta \gg 1$, then the plasma flow can be treated in the zeroth approximation in the small parameter δ^{-1} as steady-state, stationary flow:

$$\varepsilon^2 (\mathbf{v}\nabla)\,\mathbf{v} = -\frac{1}{\rho}[\mathbf{H}\,\mathrm{rot}\,\mathbf{H}]. \tag{20}$$

If, on the other hand, $\delta \ll 1$, i.e., if the plasma displacements corresponding to changes in the magnetic field are small, then we can neglect the moving term in the substantial derivative, so that the equation of motion becomes

$$\varepsilon^2 \frac{\partial \mathbf{v}}{\partial t} = -\frac{1}{\rho}[\mathbf{H}\,\mathrm{rot}\,\mathbf{H}], \tag{21}$$

while the other equations remain linear. This case corresponds to small displacements of the plasma from its equilibrium position; i.e., it corresponds to small perturbations.

In general we have $\delta \simeq 1$, and the system of MHD equations for an ideal medium takes the following form in the approximation of a strong magnetic field and a cold plasma:

$$\varepsilon^2 \frac{d\mathbf{v}}{dt} = -\frac{1}{\rho}[\mathbf{H}\,\mathrm{rot}\,\mathbf{H}], \tag{22}$$

$$\frac{\partial \mathbf{H}}{\partial t} = \mathrm{rot}\,[\mathbf{v}\mathbf{H}], \tag{23}$$

$$\frac{\partial \rho}{\partial t} + \mathrm{div}\,\rho\mathbf{v} = 0. \tag{24}$$

We seek a solution of this system as a series in the small parameter ε^2; i.e., we write all the unknowns in the form

$$f(r,\ t) = f^{(0)}(r,\ t) + \varepsilon^2 f^{(1)}(r,\ t) + \ldots \tag{25}$$

Then in zeroth order in ε^2 the magnetic field is given by

$$[\mathbf{H}^{(0)}\,\mathrm{rot}\,\mathbf{H}^{(0)}] = 0 \tag{26}$$

and by the boundary conditions, which are generally time-dependent.

As it varies over time in accordance with the boundary conditions, the magnetic field causes motion of the plasma. The kinematics of this motion is governed unambiguously by two conditions. The first follows from (22) and means that the acceleration is orthogonal to the magnetic lines of force,

$$\mathbf{H}^{(0)} \frac{d\mathbf{v}^{(0)}}{dt} = 0; \tag{27}$$

the second is a consequence of the frozen-in equation, (23),

$$\frac{\partial \mathbf{H}^{(0)}}{\partial t} = \mathrm{rot}\,[\mathbf{v}^{(0)}\mathbf{H}^{(0)}]. \tag{28}$$

Along with the continuity equation

$$\frac{\partial \rho^{(0)}}{\partial t} + \mathrm{div}\,\rho^{(0)}\mathbf{v}^{(0)} = 0 \tag{29}$$

Eqs. (26)-(28) completely determine the zeroth-order unknowns $\mathbf{H}^{(0)}(\mathbf{r}, t)$, $\mathbf{v}^{(0)}(\mathbf{r}, t)$, and $\rho^{(0)}(\mathbf{r}, t)$.

We restrict the analysis below to zeroth order in ε^2, neglecting the deviation of the magnetic field from a force-free (potential) field. However, it is not difficult to see that the subsequent application of expansion (25) to system (22)-(24) will lead to a closed system of equations for quantities of any order in ε^2.

The approximation of a strong magnetic field and a cold plasma is of considerable interest, particularly for astrophysical applications of plasma physics. Among these applications we single out two, that of the earth's magnetosphere and the solar atmosphere, as being of special importance. Common to these problems are a strong magnetic field and a low-density plasma (see, e.g., [20-23]), so that the approximation under consideration here is valid. Similar conditions are arranged in laboratory simulations of these phenomena [24-26]. We will also discuss certain other astrophysical applications of this approximation, but this paper focuses on the properties of hydrodynamic plasma flow in a strong magnetic field. Three chapters are devoted to this topic.

In the first chapter we consider continuous plasma flows in a strong magnetic field in the absence of singular null (H = 0) points.

As was shown in [19] the existence of such points leads to the appearance of current sheets. The general conditions for the appearance of current sheets in a strong magnetic field and their relationship with singular null points (or lines) were studied in [19]. Problems on the appearance of current sheets as a plasma moves in a dipole magnetic field are of interest for certain concrete applications. They are dealt with in the second chapter.

The third chapter deals with MHD plasma flow near a current sheet which has developed from a null line.

The basic results and conclusions are summarized in the Conclusion.

CHAPTER I

HYDRODYNAMIC PLASMA FLOW IN A STRONG MAGNETIC FIELD IN THE ABSENCE OF NULL LINES

§ 1. Existence of Continuous Plane (Two-Dimensional) Plasma Flows in a Strong Frozen-in Magetic Field

We begin with a discussion of two-dimensional MHD problems, which can reveal certain basic features of plasma flows in a strong frozen-in magnetic field without raising all the mathematical difficulties of the three-dimensional problems. (These results were obtained in [19] and are summarized here.) Furthermore, in certain cases two-dimensional problems turn out to give a good approximation of real three-dimensional plasma flow and in such cases can be used for a quantitative as well as qualitative comparison of theory with experiment.

A. Formalism of Two-Dimensional MHD Problems

There are two possible types of MHD problems dealing with plane (two-dimensional) plasma flows, i.e., flows with a velocity field of the type

$$\mathbf{v} = \{v_x(x, y, t),\ v_y(x, y, t),\ 0\}. \tag{1.1}$$

The first type includes problems in which the magnetic field is everywhere parallel to some direction (the z axis of a Cartesian coordinate system). The corresponding system of currents is evidently in the plane orthogonal to this direction, i.e., in the (x, y) plane. In this case all quantitities are function of x, y, and t only:

$$\mathbf{H}=\{0,\ 0,\ H(x,\ y,\ t)\},\qquad \mathbf{j}=\{j_x(x,\ y,\ t),\ j_y(x,\ y,\ t),\ 0\}. \tag{1.2}$$

We are interested in problems of the second type, involving plane plasma flows associated with a plane magnetic field. The currents corresponding to such a field are parallel to the z axis:

$$\mathbf{H}=\{H_x(x,\ y,\ t),\ H_y(x,\ y,\ t),\ 0\},\qquad \mathbf{j}=\{0,\ 0,\ j(x,\ y,\ t)\}. \tag{1.3}$$

The vector potential **A** of such a field (**H** = rot **A**) has only a single nonvanishing component (the z component):

$$\mathbf{A}=\{0,\ 0,\ A(x,\ y,\ t)\}. \tag{1.4}$$

The magnetic field intensity **H** is, by definition,

$$\mathbf{H}=\{\partial A/\partial y,\ -\partial A/\partial x,\ 0\}. \tag{1.5}$$

The function A(x, y, t), the "vector potential," is convenient because of the following properties:

1. Substitution of (1.5) into the differential equations of the lines of force,

$$\frac{dx}{H_x}=\frac{dy}{H_y}=\frac{dz}{H_z}, \tag{1.6}$$

and integration of the latter show that

$$A(x,\ y,\ t)=\text{const}\quad\text{for}\quad t=\text{const} \tag{1.7}$$

is the equation of a family of lines of force at time t in the z = const plane.

2. Let us calculate the magnetic flux through an arc element dl of contour $\mathscr{L}$ in the (x, y) plane. By definition,

$$d\Phi=\mathbf{H}[\mathbf{e}_z d\mathbf{l}]=(H_x dy-H_y dx)=dA. \tag{1.8}$$

Integrating (1.8) along contour $\mathscr{L}$ from point 1 to point 2, we find

$$\Phi=A_2-A_1. \tag{1.9}$$

Accordingly, not only is a fixed value of the vector potential A the "label" of a line of force, (1.7), but the difference between the values of A at two lines of force is equal to the magnetic flux between them. In particular, we thus find the following rule: Lines of force should be plotted in such a manner that the increment in the value of A from line to line remains constant.

3. Substituting definition (1.5) into differential equation (23), we find (within the gradient of an arbitrary function, which can be eliminated by choosing the appropriate gauge, and which is at any rate inconsequential here) the equation

$$\frac{dA}{dt}\equiv\frac{\partial A}{\partial t}+\mathbf{v}\nabla A=0. \tag{1.10}$$

This equation means that the surfaces A(x, y, t) = const are Lagrange surfaces, i.e., that they move along with the plasma. However, since, according to (1.7), these surfaces are constructed from lines of force, Eq. (1.10) describes the circumstance that the magnetic field is frozen in the plasma, as is the case in this approximation of an ideal medium. Hence we immediately find one of the integrals of motion [it can be found formally from (1.10) through a transformation to Lagrange variables].

$$A(x, y, t) = A(x_0, y_0, 0) \equiv A_0 \tag{1.11}$$

for any t. Here x_0, y_0 are the inertial coordinates of some fluid particle, and x, y are the coordinates of the same particle at time to or of any other particle which is on the same line of force A_0 at time t.

4. Written in terms of the vector potential A(x, y, t), the equation of motion (22) becomes [16]

$$\varepsilon^2 \frac{d\mathbf{v}}{dt} = -\frac{1}{\rho} \Delta A \nabla A. \tag{1.12}$$

In zeroth order in ε^2 (away from null points and magnetic field sources) we have

$$\Delta A = 0; \tag{1.13}$$

i.e., the vector potential is a harmonic function. Accordingly, with the (x, y) plane treated as the complex plane z = x + iy, it is convenient to associate with this vector potential the function

$$F(z, t) = A(x, y, t) + iB(x, y, t) \tag{1.14}$$

which is analytic in this region. Here B(x, y, t) is the conjugate harmonic function, which is related to A(x, y, t) by the Cauchy-Riemann condition [27] and is

$$B(x, y, t) = \int \left(-\frac{\partial A}{\partial y} dx + \frac{\partial A}{\partial x} dy\right) + B_0(t) = -\int \mathbf{H} d\mathbf{l} + B_0(t), \tag{1.15}$$

where $B_0(t)$ is a quantity independent of x and y.

We call the function F(z, t) the "complex potential." According to Eqs. (1.5) and (1.14) the magnetic field vector is

$$\mathbf{H} = H_x + iH_y = -i(\partial F/\partial z)^*, \tag{1.16}$$

where the asterisk denotes complex conjugation.

As we mentioned in the Introduction, the kinematics of a plasma associated with the changes in a potential magnetic field is unambiguously governed by two conditions in the approximation of a strong field and a cold plasma: the condition that the acceleration be orthogonal to the magnetic lines of force,

$$\left[\frac{\partial \mathbf{v}}{\partial t} \nabla A\right] = 0, \tag{1.17}$$

and the frozen-in condition, (1.10). We note for clarity that Eq. (1.17) is the result of eliminating the unknown $\Delta A^{(1)}$, which is of first order in ε^2, from the two components of the vector equation

$$\frac{d\mathbf{v}^{(0)}}{dt} = -\frac{1}{\rho^{(0)}} \Delta A^{(1)} \nabla A^{(0)}. \tag{1.18}$$

B. Existence of Continuous Solutions [19]

Accordingly, in the approximation of a strong magnetic field and a cold plasma the MHD equations for plane two-dimensional flow of an ideally conducting plasma reduce in zeroth order to the following system of equations [16-19]:

$$\Delta A = 0, \tag{1.19}$$

$$\left[\frac{d\mathbf{v}}{dt} \nabla A\right] = 0, \tag{1.20}$$

$$\frac{dA}{dt} = 0, \tag{1.21}$$

$$\frac{\partial \rho}{\partial t} + \operatorname{div} \rho \mathbf{v} = 0. \tag{1.22}$$

It would seem that the solution of this system is unambiguously defined within some region G on the (x, y) plane if the boundary conditions are specified at boundary S,

$$A(x, y, t)|_S = f_1(x, y, t), \tag{1.23}$$

and if the initial conditions are specified within region G,

$$\mathbf{v}_{\parallel}(x, y, 0) = \mathbf{f}_2(x, y), \tag{1.24}$$

$$\rho(x, y, 0) = f_3(x, y). \tag{1.25}$$

Here $\mathbf{v}_{\parallel}$ is the velocity component along the lines of force. The velocity component across the lines of force is unambiguously governed by Eq. (1.21) for a known potential A(x, y, t) and is equal at all times (including the initial time) to

$$\mathbf{v}_{\perp} = (\mathbf{v}\nabla A)\nabla A/|\nabla A|^2 = -\frac{\partial A}{\partial t}\nabla A/|\nabla A|^2. \tag{1.26}$$

At any time we can find A(x, y, t) from Eq. (1.19) and boundary condition (1.23). Then from Eqs. (1.20) and (1.21) and initial condition (1.24) we can determine the velocity $\mathbf{v}(x, y, t)$; from the continuity equation and condition (1.25) we find $\rho(x, y, t)$.

However, it is by no means in all cases that continuous solutions exist for system (1.19)-(1.21), which describe the continuous deformation of the magnetic field and the corresponding continuous plasma flows [19]. For example, if the boundary conditions and initial conditions are specified in such a manner that they do not violate the original assumptions regarding the small parameters γ^2 and ε^2, i.e., inequalities (19), then the vector potential A(x, y, t) is unambiguously determined by Eq. (1.19) and boundary condition (1.23). However, this latter condition can in general be such that the magnetic field corresponding to potential A contains null points ($H = |\nabla A| = 0$), including null points for which the electric field $E = -c^{-1}\partial A/\partial t$ is nonvanishing. Such points contradict the frozen-in equation (1.21). Following [19], we call such null points "singular."

Accordingly, the frozen-in condition permits continuous deformation of the strong magnetic field and the corresponding continuous plasma motion everywhere except at singular null points (lines parallel to the z axis of the Cartesian coordinate system) of the magnetic field, at which we have

$$\mathbf{H} = \operatorname{rot} \mathbf{A} = 0, \quad \mathbf{E} = -\frac{1}{c}\frac{\partial \mathbf{A}}{\partial t} \neq 0. \tag{1.27}$$

In the present chapter we consider some simple examples of plasma flow in strong magnetic fields without singular null points. As the first example we treat plasma motion in the field of a plane (two-dimensional) magnetic dipole which varies with the time.

C. Plasma Motion in the Field of a Plane Magnetic Dipole Which Varies with the Time

Two parallel straight currents identical in magnitude but opposite in direction produce a magnetic field which can be described, sufficiently far from the currents, by a potential

$$F(z)=\frac{i\mathbf{m}}{z}, \qquad \mathbf{m}=me^{i\psi}. \tag{1.28}$$

Such a field is called a "plane dipole magnetic field" or the "field of a plane magnetic dipole." The quantity $m=2\mathcal{J}l/c$ represents the dipole moment, $\mathcal{J}$ is the magnitude of the currents, and l is the distance between the currents. Equation (1.28) corresponds to a plane magnetic dipole at the origin in the (x, y) plane, oriented at an angle ψ with respect to the x axis (the currents are parallel to the z axis of the Cartesian coordinate system).

We consider the plasma flow due to a change over time in the strong magnetic field of a plane dipole frozen in the plasma. We assume $\psi = \pi/2$, and $m = m(t)$.

The components of the magnetic field intensity of this field are

$$\mathbf{H}=\left\{\frac{2mxy}{(x^2+y^2)^2},\quad \frac{m(y^2-x^2)}{(x^2+y^2)^2},\ 0\right\}, \tag{1.29}$$

and the lines of force are, according to (1.7) and (1.28), the family of circles

$$m\frac{x}{x^2+y^2}=\text{const} \tag{1.30}$$

which are centered on the x axis and have the common point $x = 0$, $y = 0$.

Correspondingly, frozen-in condition (1.11) leads to

$$m\frac{x}{x^2+y^2}=m_0\frac{x_0}{x_0^2+y_0^2}. \tag{1.31}$$

Here x_0, y_0 are the coordinates of some "fluid particle" at the initial time ($t = 0$), and x, y are the coordinates of the same particle at time t. The dipole moment is treated as a function of the time: $m = m(t)$, with $m_0 = m(0)$.

The second integral of motion can be found easily by restricting the analysis to small changes in the dipole moment m(t) and correspondingly small plasma displacements. Assuming the parameter $\delta = VT/L$ to be small, we replace Eq. (1.20) by linear equation (21), which takes the following form, after an integration over time with a zero initial velocity, for the case of a plane dipole field which retains its spatial shape:

$$\frac{dx}{dt}=K(x,\ y,\ t)\frac{\partial A}{\partial x}, \qquad \frac{dy}{dt}=K(x,\ y,\ t)\frac{\partial A}{\partial y}. \tag{1.32}$$

Here K(x, y, t) is some unknown function of the coordinates and the time; eliminating it from Eqs. (1.32), we find the differential equation

$$\frac{dy}{dx}=\left(\frac{\partial A}{\partial y}\Big/\frac{\partial A}{\partial x}\right). \tag{1.33}$$

We see that in this approximation not only the acceleration but also the corresponding plasma displacements are everywhere directed along the normal to lines of force.

Substituting (1.29) into (1.33) we find the simple differential equation

$$\frac{dy}{dx}=\frac{2xy}{x^2-y^2}. \tag{1.34}$$

Its integral

$$\frac{y}{x^2+y^2}=\text{const} \tag{1.35}$$

is the equation of a family of circles orthogonal to lines of force (1.30) and describes the trajectories of the fluid particles.

In particular, the trajectory of a fluid particle initially (at t = 0) at the point (x_0, y_0) is the arc of the circle

$$\frac{y}{x^2+y^2}=\frac{y_0}{x_0^2+y_0^2} \tag{1.36}$$

from the point (x_0, y_0) to the point (x, y) along line of force (1.31).

Accordingly, the integrals in (1.31) and (1.36) completely determine the plasma flow in terms of Lagrange coordinates:

$$x=x(x_0,\ y_0,\ t), \qquad y=y(x_0,\ y_0,\ t). \tag{1.37}$$

This flow is simple in nature: The fluid particles are bound to lines of force and move along with them in the transverse direction. This simple kinematics is of course the result of our assumption of small displacements of the plasma from a state with a zero initial velocity under the influence of a force perpendicular to the magnetic lines of force.

The change in the density of the fluid particle as it moves along the trajectory found here is governed by the continuity equation (1.22) written in Lagrange form:

$$\frac{\rho(x,\ y,\ t)}{\rho_0(x_0,\ y_0)}=\frac{d\Omega_0}{d\Omega}=\frac{\mathcal{D}(x_0,\ y_0)}{\mathcal{D}(x,\ y)}, \tag{1.38}$$

where $d\Omega_0$ is the initial volume of the fluid particle, $d\Omega$ is the volume of the same particle at time t, and $\mathcal{D}(x_0, y_0)/\mathcal{D}(x, y)$ is the Jacobian of the transformation inverse to (1.37) for a fixed value of t.

In the particular case of a homogeneous initial distribution of the plasma density,

$$\rho_0(x_0,\ y_0)\equiv\rho_0=\text{const}, \tag{1.39}$$

Eq. (1.38) gives the plasma distribution in the (x, y) plane at time t.

In general the plasma-density distribution in the (x, y) plane is

$$\rho(x,\ y,\ t)=\rho_0[x_0(x,\ y,\ t),\ y_0(x,\ y,\ t)]\frac{\mathcal{D}[x_0(x,\ y,\ t),\ y_0(x,\ y,\ t)]}{\mathcal{D}(x,\ y)}. \tag{1.40}$$

Calculating the Jacobian of (1.38) for transformation (1.31), (1.36), we find

$$\frac{\rho(x,\ y,\ t)}{\rho_0}=\left(\frac{m}{m_0}\right)\frac{m_0^4\{[m^2x^4+m_0^2y^4+x^2y^2(3m^2-m_0^2)]^2-[2x^2y^2(m_0^2-m^2)]^2\}}{(m^2x^2+m_0^2y^2)^4}. \tag{1.41}$$

In particular, at the dipole axis (x = 0) we have

$$\frac{\rho(0,\ y,\ t)}{\rho_0}=\frac{m}{m_0}, \tag{1.42}$$

while in the "equatorial" plane (y = 0) we have

$$\frac{\rho(x,\ 0,\ t)}{\rho_0}=\left(\frac{m_0}{m}\right)^3. \tag{1.43}$$

Accordingly, as the dipole moment m increases, the plasma density at the dipole axis increases in proportion, while that in the equatorial plane falls off in inverse proportion to the cube of the dipole moment. As the dipole moment is reduced, of course, these densities change in the opposite direction.

We recall that this result holds for the case of small changes in the dipole moment and can thus at best demonstrate a tendency in the behavior of a plasma in a strong magnetic field of a varying plane dipole. An exception to this statement is Eq. (1.42) which holds for any changes in the dipole moment. As we mentioned above, in the approximation of a strong magnetic field and a cold plasma the acceleration is perpendicular to lines of force. Accordingly, if the plasma was initially at rest, motion does not appear at the dipole axis, regardless of the changes in the dipole moment, while near the axis the plasma displacements are always small ($\delta \ll 1$), and the solution found here is valid.

In the general case of arbitrarily large changes in the dipole moment, to solve this problem rigorously we must integrate Eq. (1.20) along with Eq. (1.21). Here inertial effects, which lead to plasma motion along the lines of force, are important. The solution of this type for plasma motion in the magnetic field of a three-dimensional (axisymmetric) dipole which is growing stronger is discussed in the next section.

§ 2. Continuous Plasma Flows in a Strong, Poloidal, Axisymmetric Magnetic Field

The problem of the plasma motion in a strong, poloidal ($H_\varphi = 0$), axisymmetric magnetic field was formulated in [18] and solved for a dipole magnetic field in the approximation of small changes in the dipole moment and, correspondingly, small plasma displacements ($\delta \ll 1$). An analytic solution was found for the linearized problem. This solution gives the direction in which the various quantities change; in particular, it leads to a conclusion regarding the plasma density near the axis of a growing magnetic dipole which is important in applications.

For concrete applications in astrophysics and the physics of laboratory plasmas it is important to have the exact nonlinear solutions corresponding to arbitrary (not small) changes in the magnetic moment, in which cases we can expect the appearance of dense plasma blobs or, in astrophysical terms, plasma condensations.

In this section we derive the corresponding nonlinear equations; they are analyzed and solved numerically in detail in [28].

A. General Formulation of the Problem

We assume that the plasma motion is due to the axisymmetric magnetic field of some central object and is described by the MHD equations for an ideal compressible medium. These equations are written in terms of dimensionless quantities and spherical coordinates as follows (the axial symmetry is taken into account):

$$\frac{\varepsilon^2}{\delta}\frac{\partial \mathbf{v}}{\partial t}+\varepsilon^2(\mathbf{v}\nabla)\mathbf{v}=-\gamma^2\frac{\nabla p}{\rho}+\frac{j_\varphi}{\rho r\sin\theta}\nabla\Phi, \tag{1.44}$$

$$\frac{\partial \Phi}{\partial t}+\delta\mathbf{v}\nabla\Phi=0, \tag{1.45}$$

$$\frac{\partial \rho}{\partial t} + \delta \operatorname{div} \rho v = 0. \tag{1.46}$$

Here $\Phi = \Phi(r, \theta, t)$ is the dimensionless "stream function," which is related by definition to the only nonvanishing (φ) component of the vector potential **A** by

$$\Phi(r, \theta, t) = r \sin\theta A_\varphi(r, \theta, t); \tag{1.47}$$

the quantity

$$j_\varphi = -\frac{1}{r}\left[\frac{1}{\sin\theta}\frac{\partial^2\Phi}{\partial r^2} + \frac{\partial}{\partial\theta}\left(\frac{1}{r^2\sin\theta}\frac{\partial\Phi}{\partial\theta}\right)\right] \tag{1.48}$$

is the φ component of the dimensionless current density **j**; and ε, δ, and γ are the dimensionless parameters in (13).

By definition the components of the magnetic field corresponding to (1.47) are

$$H_r = \frac{1}{r\sin\theta}\frac{\partial}{\partial\theta}(\sin\theta A_\varphi) = \frac{1}{r^2\sin\theta}\frac{\partial\Phi}{\partial\theta},$$
$$H_\theta = -\frac{1}{r}\frac{\partial}{\partial r}(rA_\varphi) = -\frac{1}{r\sin\theta}\frac{\partial\Phi}{\partial r}. \tag{1.49}$$

As in [18] we use the approximation of a strong magnetic field and a cold plasma, (19), but, in contrast with [18], we do not assume that the changes in the magnetic field and the corresponding plasma motion are small. On the contrary, we set $\delta = 1$; we will see below that this choice determines the unit of time t.

Under these assumptions the original system of equations becomes

$$\varepsilon^2 d\mathbf{v}/dt = K(r, \theta, t)\nabla\Phi, \tag{1.50}$$

$$d\Phi/dt = 0, \tag{1.51}$$

$$d\rho/dt = -\rho \operatorname{div} \mathbf{v}, \tag{1.52}$$

where $K(r, \theta, t) = j_\varphi(r, \theta, t)/\rho r \sin\theta$.

We seek a solution of this system of equations as a series in the small parameter ε^2:

$$\Phi(r, \theta, t) = \Phi^{(0)}(r, \theta, t) + \varepsilon^2\Phi^{(1)}(r, \theta, t) + \ldots \tag{1.53}$$

In the zeroth approximation in ε^2 we have $K^{(0)}(r, \theta, t) = 0$, i.e.,

$$j_\varphi^{(0)}(r, \theta, t) = 0, \tag{1.54}$$

corresponding to a time-evolving potential magnetic field described by the stream function $\Phi^{(0)}(r, \theta, t)$.

From the equation of motion (1.50) we find

$$d\mathbf{v}^{(0)}/dt = K^{(1)}(r, \theta, t)\nabla\Phi^{(0)}. \tag{1.55}$$

Accordingly, in zeroth order in ε^2 the acceleration is perpendicular to the lines of force of the potential magnetic field. Equations (1.54) [see (1.48), (1.55), and (1.51)] completely determine the zeroth-order unknowns $\Phi^{(0)}$ and $\mathbf{v}^{(0)}$. We note that these equations, written in terms of the stream function, are formally the same as the corresponding equations describing plane plasma flow in terms of the vector potential A, Eqs. (1.19)-(1.21). We can therefore generalize the condition for the existence of continuous solutions which was found in [19] for plane flows to the case of plasma flow in a strong, axisymmetric, poloidal magnetic field. Specifically, there can be a continuous deformation of a strong axisymmetric poloidal magnetic field and the corresponding continuous meridional motion of the plasma everywhere except at singular null points of the magnetic field, i.e., at points at which

$$H_r = H_\theta = 0, \quad E_\varphi = -\frac{1}{c}\frac{\partial}{\partial t}\left(\frac{\Phi(r, \theta, t)}{r \sin\theta}\right) \neq 0. \tag{1.56}$$

We thus have grounds for assuming the existence of a continuous plasma flow in a strong magnetic field of the dipole type in which there are clearly no such points.

B. Case of a Dipole Magnetic Field

As the zeroth-approximation magnetic field we adopt a dipole field, for which the stream function is

$$\Phi^{(0)}(r, \theta, t) = m(t)\sin^2\theta/r, \tag{1.57}$$

where m(t) is the time-evolving magnetic moment. For example, in the case of a uniformly magnetized gaseous sphere (a star) of radius R(t) with a frozen-in internal magnetic field $H_i(t)$ we would have

$$m(t) = \frac{1}{2} H_i(t) R^3(t) = \frac{1}{2} H_0 R_0^2 R(t), \tag{1.58}$$

where H_0 and R_0 are the initial values of $H_i(t)$ and R(t) (at time t = 0).

The stream function $\Phi(r, \theta, t)$ is convenient because, as the formal analog of the vector potential A(x, y, t) in plane problems, it has all the properties of this vector potential; in particular, as can be seen from Eq. (1.51), it is an integral of motion:

$$\Phi(r, \theta, t) = \Phi(r_0, \theta_0, 0). \tag{1.59}$$

Here the subscript "0" denotes the initial values of the Lagrange coordinates $r = r(r_0, \theta_0, t)$, $\theta = \theta(r_0, \theta_0, t)$ of the fluid particle. Accordingly, for the dipole field we have the first integral

$$\frac{m(t)\sin^2\theta}{r} = \frac{m_0 \sin^2\theta_0}{r_0}. \tag{1.60}$$

Transforming to the Lagrange coordinates r and θ in Eq. (1.55) and substituting the stream function (1.57) into (1.55) we find

$$\ddot{r} - r\dot{\theta}^2 = -\frac{Km}{r^2}\sin^2\theta, \tag{1.61}$$

$$r\ddot{\theta} + 2\dot{r}\dot{\theta} = \frac{Km}{r^2} 2\sin\theta\cos\theta. \tag{1.62}$$

Dividing the radial component of the equation of motion in (1.61) by the angular component in (1.62), and using (1.60), we can eliminate the function $r = r(r_0, \theta_0, t)$. The result is an ordinary differential equation for the function $\theta = \theta(r_0, \theta_0, t)$:

$$ma(\theta)\ddot{\theta} + mb(\theta)\dot{\theta}^2 + 2\dot{m}a(\theta)\dot{\theta} + \ddot{m}c(\theta) = 0, \tag{1.63}$$

where

$$\begin{aligned} a(\theta) &= \sin\theta(3\cos^2\theta + 1), \\ b(\theta) &= 2\cos\theta(3\cos^2\theta - 1), \\ c(\theta) &= 2\sin^2\theta\cos\theta. \end{aligned} \tag{1.64}$$

Here a dot denotes differentiation with respect to the time.

We assume that the gas is initially at rest so that the solution $\theta(t)$ of Eq. (1.63) satisfies the initial conditions

$$\theta(0)=\theta_0, \qquad \dot{\theta}(0)=\dot{\theta}_0=0. \tag{1.65}$$

We note that the solution of Eq. (1.63) is independent of r_0. The function $r = r(r_0, \theta_0, t)$ is determined from the solution $\theta = \theta(t, \theta_0)$ with the help of (1.60):

$$r(r_0, \theta_0, t)=r_0\frac{m(t)\sin^2\theta(t, \theta_0)}{m_0\sin^2\theta_0}. \tag{1.66}$$

This similarity property of the solution with respect to the initial radius vector r_0 of the fluid particle is a consequence of the simple geometry of the magnetic field: The shape of the lines of force of the dipole, which is changing in magnitude, is the same at all times.

The meaning of this similarity can also be understood easily on the basis of the theory of dimensionality. Of the five independent dimensional parameters ρ, m, r, θ, t, the first three have independent dimensionalities; accordingly, it is possible to construct only two dimensionless combinations, which are the independent variables. One is the polar angle θ; it is convenient to choose the dimensionless time as the second.

Accordingly, if the fluid particle is initially at rest at the point r_0, θ_0, then, as we see from (1.66), its coordinates and velocity components at time t are given by

$$\theta=\theta(t, \theta_0), \tag{1.67}$$

$$r=r_0\mathcal{R}(t, \theta), \qquad v_r=r_0\mathcal{V}_r(t, \theta), \qquad v_\theta=r_0\mathcal{V}_\theta(t, \theta), \tag{1.68}$$

where

$$\begin{gathered}\mathcal{R}(t, \theta)=\frac{m(t)\sin^2\theta}{m_0\sin^2\theta_0},\\ \mathcal{V}_r(t, \theta)=\frac{\partial\mathcal{R}}{\partial t}+\frac{\partial\mathcal{R}}{\partial\theta}\dot{\theta}, \qquad \mathcal{V}_\theta(t, \theta)=\dot{\theta}\mathcal{R}(t, \theta),\end{gathered} \tag{1.69}$$

and θ is the running angular coordinate defined in (1.67). In general the function $\mathcal{R}(t, \theta)$ and its derivatives depend on θ_0 not only through $\theta(t, \theta_0)$ but also directly, as can be seen from the definition. However, for simplicity we will not write this part of the argument. In particular, for a density change we have

$$\rho(r, \theta, \varphi, t)=\rho_0(r_0, \theta_0, \varphi_0)\mathcal{P}(t, \theta), \tag{1.70}$$

where $\rho_0(r_0, \theta_0, \varphi_0)$ is the initial profile of the plasma density and

$$\mathcal{P}(t, \theta)=\frac{r_0^2\sin\theta_0\mathcal{D}(r_0, \theta_0)}{r^2\sin\theta\mathcal{D}(r, \theta)}=\frac{r_0^2\sin\theta_0}{r^2\sin\theta}\frac{\partial r_0}{\partial r}\frac{\partial\theta_0}{\partial\theta}=\left(\frac{m_0}{m(t)}\right)^3\frac{\sin^7\theta_0(\theta, t)}{\sin^7\theta}\frac{\partial\theta_0(\theta, t)}{\partial\theta}. \tag{1.71}$$

In deriving Eq. (1.71) we took into account the similarity property of the solution and Eq. (1.60). By $\theta_0(\theta, t)$ here we understand the function which is the inverse of the solution $\theta = \theta(t, \theta_0)$ of Eq. (1.63) for a given value of t.

If the initial density distribution is uniform, $\rho_0(r_0, \theta_0, \varphi_0) \equiv \rho_0 = \text{const}$, then we have, according to (1.70),

$$\rho=\rho(\theta, t)=\rho_0\mathcal{P}(t, \theta). \tag{1.72}$$

Accordingly, with ρ_0 = const, the distribution at time t is independent of the radius r, so that the constant-density surfaces are the conical surfaces θ = const. This feature of the problem is also a consequence of the similarity property discussed above.

The solution found for the problem in [28] shows that as the dipole moment increases (as the sphere expands) the magnetic field "rakes" the plasma toward the dipole axis, compresses it, and simultaneously accelerates it along the lines of force. The plasma density at the dipole axis, as in the two-dimensional case, (1.42), typically increases in proportion to the dipole moment.

The solution is applicable in a certain axisymmetric shell. The inner boundary of this shell is governed by the size of the region outside which the field can be assumed a dipole field. The existence of the outer boundary and the shape of this boundary are governed by the circumstance that the assumption that the parameters γ^2 and ε^2 are small breaks down at certain distances from the dipole. For example, the magnetic energy density, falling off as r^{-6}, becomes equal to the initial thermal energy density of the plasma n_0kT_0 at some distance R_1; equivalently, the Alfvén velocity becomes equal to the sound velocity. At the dipole axis, where the plasma density is higher because of the "raking," the parameter γ^2 becomes equal to unity at $r = R_2 < R_1$. Another limitation in the equatorial region results from the increase of the plasma velocity with distance from the dipole; as a result, the condition $\varepsilon^2 \ll 1$ breaks down at some point. Accordingly, the application of the resulting solution to concrete problems involves the satisfaction of stringent conditions, which ultimately reduce to the requirement that the magnetic field be strong.

In astrophysical situations, conditions of this type can be satisfied in stellar atmospheres (and in planets) where the energy density of the magnetic field produced by the processes within the star (or planet) can be large in comparison with the internal-energy density of the low-density plasma atmosphere. In this case the plasma behavior is controlled entirely by the magnetic field, so that the dynamics is unusual. At the sun, for example, this situation is manifested by the formation of coronal condensations, ejections, spicules, certain types of protuberances, and other structural elements of the chromosphere and corona.

Analogous processes, but on a much larger scale, should operate in the atmospheres of magnetically variable stars and novae, if the latter have strong magnetic fields. For example, in the explosion of a magnetic star the growing magnetic moment of the expanding envelope should lead, according to the solution found above, to condensation of the surrounding low-density gas near the magnetic poles of the star. This process could be one reason for the observed regular structure of the envelopes of novae [29, 30].

In the laboratory this magnetic raking can be used to produce dense hot plasmas. This process is essentially a version of the known process of magnetic plasma compression [31], with the distinguishing feature that the gas-pressure effect is negligibly small.

The next two sections of this chapter deal with astrophysical applications of these solutions.

§ 3. Plowing of the Interstellar Medium by the Magnetic Field of an Expanding Envelope

A. Observation of Gaseous Condensations in the Envelopes of Novae

The envelopes of novae [32, 29] and supernovae [33] as well as planetary nebulae [34-36] are known to take on a wide variety of shapes. Even among envelopes of regular shape, perfect roundness is rare; it is more common to find oblate or prolate axisymmetric envelopes. As a rule the surface brightness of such envelopes reaches maxima at the ends of one of the

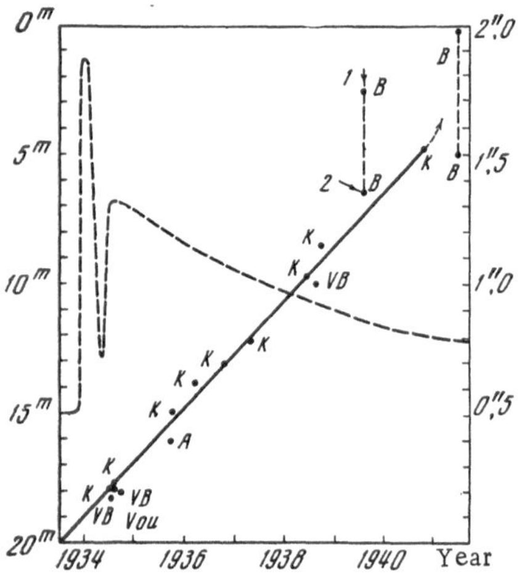

Fig. 1. Brightness curve of the nova DQ Her 1934 (dashed curve), angular distance between condensations (straight line), and angular dimension of the major (1) and minor (2) semi-axes of the envelope. Observer and reference: A) Aitken [39]; B) Baade [43]; K) Kuiper [37, 38]; VB) Van Biesbroeck [41, 42]; Vou) Voute [40].

principal axes of the oval observed image. In certain cases this phenomenon can be interpreted as a glowing gas ring observed nearly in its plane. However, if there is no bright band typical of a ring between the brightness maxima, we are left with the assumption that there are two distinct gaseous condensations in the envelope.

In a nova explosion, in the early stage of the envelope expansion, the gaseous condensations sometimes become so bright that they give the impression that the nova "splits" at the time of the explosion. For example, an apparent splitting of the nova DQ Her 1934 was observed for several years [37-42]. Figure 1 shows the brightness curve for DQ Her 1934, the angular envelope "radius," and the angular distance between the condensations.

In examining this figure the reader should bear in mind that the points from 1935 through 1939 were obtained by visual micrometric observations carried out at the limiting resolution of the corresponding telescopes, so the corresponding data are not very reliable. Nevertheless, the general behavior of these points and the photographic measurements since 1939 make it extremely probable that symmetric gaseous condensations do exist (we will call them "polar condensations"), expanding in opposite directions at a constant velocity.

The polar condensations became noticeable half a year after the explosion. Comparison of the angular velocity of the envelope expansion (~0.5 arc second per year) [43, 30] with the angular velocity of the outward motion of the condensation (~0.2 arc second per year according to Fig. 1) shows that the envelope expanded roughly twice as fast as the condensations moved outward. With a large body of observational evidence available on the Doppler shift of the emission lines [44-47], it is possible to place the expansion velocity of the envelope at about 300 km/sec, and the distance to the nova at 300 pc. The condensations move at about 150 km/sec; because of this difference in velocities the envelope overtakes the polar condensations, so that the photographs obtained in 1942 [43] do not reveal the individual condensations—only slight inhomogeneities near the polar regions of the envelope.

The nova DQ Her 1934 does not represent the only known case of polar condensations; analogous condensations have been found for V 603 Aql 1918 [48-50] and certain other novae and supernovae [32, 33].

B. Theoretical Models

Here we will not take up the "nonmagnetic" models for the formation of gaseous condensation, e.g., the model relating the envelope structure to a resonance of tidal forces in a close binary system [51]; this model is discussed critically in [30].

There is considerable interest in the "semimagnetic" models, in which the magnetic field affects the envelope shape only indirectly. For example, it is known that a spherical shock wave moving outward toward the surface of an oblate, polytropic gaseous sphere can lead to the formation of condensations in the ejected envelope [52]. In this model the magnetic field can serve as the agent causing the oblate shape.

In an attempt to explain this phenomenon Mustel' [29] advanced the hypothesis that novae have strong magnetic fields. According to this hypothesis the polar condensations result from an interaction between the envelope ejected during the explosion and the magnetic field of the star.

Below we take up an alternative explanation, while retaining the concept that the condensations are of a magnetic nature; specifically, we assume that the polar condensations arise from the interstellar plasma as a result of the raking of this plasma by the magnetic field of the envelope. We assume that the star has a dipole magnetic field. The expansion of the envelope of the nova leads to an increase in its magnetic dipole moment and to a displacement of the surrounding plasma envelope toward the dipole axis; these events may be responsible for the appearance of polar condensations.

In this model we can arbitrarily divide the process of the formation of the polar condensations into two stages; in the first stage the interstellar plasma is raked by the magnetic field toward the polar regions, and the pressure and density at the dipole axis increase. In the second stage the elevated gas pressure prevents a further increase in the plasma density at the polar axis and stops the compression, but the raking of the plasma continues. At the same time, the gas-pressure gradient which arises ahead of the envelope leads to plasma motion along the polar axis, so that all the plasma is raked into two compact condensations by the time the magnetic forces become ineffective.

The interaction of these condensations with the envelope and their cooling due to expansion and radiation lead to a gradual disappearance of the bipolar structure.

Before we turn to a discussion of these stages we note that this model is fundamentally different from that proposed by Kolesnik [53]. As was mentioned in the Introduction, the magnetic field can be force-free (in particular, potential) not only when it is strong but also if nonmagnetic forces are balanced. Kolesnik [53] chose a self-similar regime for the slow expansion of a conducting medium (the envelope, according to [29]) in which a dipole magnetic field is frozen such that the inertial force precisely balances the gas-pressure gradient. As a result the magnetic field remains a dipole potential field forever.

C. Raking of the Interstellar Plasma by the Magnetic Field of the Expanding Envelope

We assume that the magnetic moment of the envelope formed during the explosion of the nova is described by Eq. (1.58), where $R(t)$ is the envelope radius, R_0 is the initial radius (the radius of the exploding star), and H_0 is the magnetic field at the surface of this star. Here we are assuming that the magnetic field in the envelope is completely entrained by the gas, i.e., that

$$\frac{H^2}{8\pi} < \frac{\rho V^2}{2}. \tag{1.73}$$

Let us check inequality (1.73) for the case of the envelope of DQ Her 1934. According to the 1968 data [30] the average density of the envelope is $\rho \sim 3 \cdot 10^{-21}$ g/cm^3 at a radius of $R \sim 3 \cdot 10^{16}$ cm. Over the time which has elapsed since the explosion, $t \sim 10^9$ sec, the envelope radius has increased by a factor of about 10^6. We can therefore estimate the average density of the envelope to have been $\bar{\rho}_0 = \bar{\rho} R^3/R_0^3 \sim 3 \cdot 10^{-3}$ g/cm^3 immediately after the explosion. With an expansion velocity of $\bar{V} \sim 3 \cdot 10^7$ cm/sec for the envelope the average kinetic energy density of the envelope at the time it was ejected was about $\bar{\rho}_0\bar{V}^2/2 \sim 1.5 \cdot 10^{12}$ ergs/cm^3. Accordingly, a field $H_0 < 6 \cdot 10^6$ G is entrained by the expanding envelope.

We note that in the case of inertial expansion of the envelope inequality (1.73) becomes stronger as time elapses, since $\bar{\rho}\bar{V}^2/2 \sim R^{-3}$ but $H^2/8\pi \sim R^{-4}$. Accordingly, the magnetic raking continues until the condition

$$\frac{H^2}{8\pi} > nkT \tag{1.74}$$

is no longer satisfied ahead of the envelope. This condition establishes the outer boundary of the raking region, whose inner boundary is the surface of the expanding envelope. We define the radius $R_1(t)$ of the outer boundary as the distance at which we have $H^2/8\pi \simeq nkT$ [cf. (1.73)]. The magnetic field within the expanding envelope of radius $R(t)$ is

$$H_{en} = H_0\left(\frac{R_0}{R}\right)^2, \tag{1.75}$$

while the field outside the envelope is

$$H = H_{en}\left(\frac{R}{r}\right)^3 = H_0\left(\frac{R}{R_0}\right)\left(\frac{R_0}{r}\right)^3; \tag{1.76}$$

accordingly, the outer radius of the magnetic-raking region is

$$R_1(t) = \left(\frac{H_0^2}{8\pi nkT} R_0^4 R^2\right)^{1/6}. \tag{1.77}$$

This raking region vanishes, and the condensation process ends, when the envelope radius R becomes equal to the outer radius in (1.77), i.e., at

$$R = R_{1\max} = R_0\left(\frac{H_0^2}{8\pi nkT}\right)^{1/4}. \tag{1.78}$$

With further expansion the influence of the magnetic field of the envelope on the surrounding interstellar gas is inconsequential.

In the estimate in (1.77) we neglected the change in the gas pressure during the magnetic raking. Now taking this change into account, we find that inequality (1.74) is violated at the dipole axis at much smaller values of R. We introduce the radius $R_2(t)$ defined by

$$\frac{H_0^2}{8\pi}\left(\frac{R}{R_0}\right)^2\left(\frac{R_0}{R_2}\right)^6 = p' = n'kT', \tag{1.79}$$

where $p' = n'kT'$ is the pressure at the dipole axis in the magnetic-compression region (the initial pressure is $p = nkT$). If the compression occurs adiabatically $[p'/p = (\rho'/\rho)^\gamma]$, we find, noting that we have $\rho'/\rho = m/m_0$ at the dipole axis,

$$\frac{R_2(t)}{R_0} = \left(\frac{H_0^2}{8\pi nkT}\right)^{1/6}\left(\frac{R(t)}{R_0}\right)^{(2-\gamma)/6}. \tag{1.80}$$

In the case of isothermal compression we would have

$$\frac{R_2(t)}{R_0} = \left(\frac{H_0^2}{8\pi nkT}\right)^{1/6}\left(\frac{R(t)}{R_0}\right)^{1/6}. \tag{1.81}$$

To choose between these two possibilities we must compare the scale time for the compression with the scale time for cooling. Simple estimates for DQ Her 1934 speak in favor of isothermal compression. In this case the compression at the dipole axis is greatest when $R = R_2$, and this compression is found from (1.81) to be

$$\frac{\rho_{max}}{\rho_0} = \left(\frac{H_0^2}{8\pi nkT}\right)^{1/5}. \tag{1.82}$$

The subsequent raking of the interstellar plasma (out to $R = R_{1max}$) by the magnetic field does not increase the density at the dipole axis. The increase in the mass of the gas being raked is accompanied by an expansion of the boundary θ_{co} of the compression region. Figure 2 shows schematic density profiles for various values of R.

What are the typical plasma velocities in the compression region? As was mentioned earlier, in the approximation of a strong field and a cold plasma there is no motion at the dipole axis because the gas-pressure gradient is neglected. However, the gas pressure in the compression region becomes important even at $R = R_2$; the gas-pressure gradient perpendicular to the lines of force stops the plasma compression, but the gradient parallel to the lines of force leads to plasma motion at the dipole axis, in the same direction as the motion of the envelope, but at a lower velocity. Neglecting ρ_0 and p_0 in comparison with ρ_{max} and p_{max}, and choosing $R_{1max}/2$ as a typical distance, we can easily estimate the hydrodynamic acceleration $\dot{v}_r$ and velocity v_r of the plasma along the dipole axis. Assuming $H_0 \sim 10^3$ G, $T_0 \sim 10^4$ °K, $n_0 \sim 2$ cm^{-3}, and an envelope velocity V ~ 300 km/sec, we find $v_r \sim 100$ km/sec. This agrees in order of magnitude with the kinematic velocity (~150 km/sec) of a fluid particle at the boundary of the compression region, at the point $r = R_{1max}$, $\theta = \theta_{co}$, at the time at which the magnetic forces become ineffective.

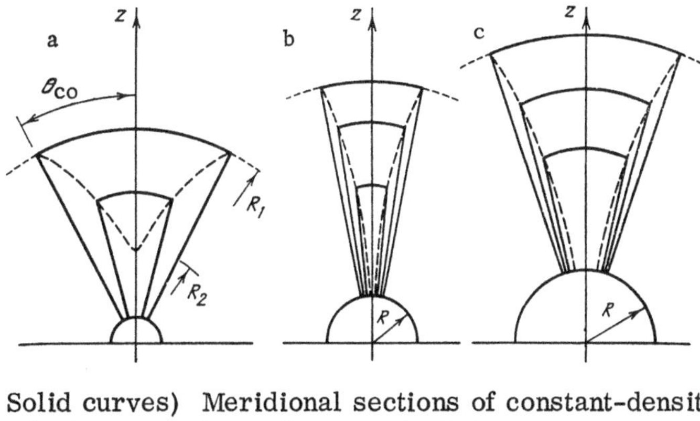

Fig. 2. Solid curves) Meridional sections of constant-density surfaces; dashed curves) the surface at which the magnetic-energy density is equal to the thermal-energy density, i.e., the boundary of the region in which the solution is applicable. a) $R < R_2 < R_1$. "Raking" of the plasma by the magnetic field increases the density at the polar axis; b) $R = R_2 < R_1$. The increased gas pressure stops the compression at the axis, and the density and pressure reach their maxima; c) the raking of the plasma continues at a constant density at the dipole axis. The increase in the mass of the polar condensations is accompanied by the expansion of the compression region, i.e., an increase of θ_{co}.

We thus find a natural explanation for the difference between the envelope expansion velocity and the velocity at which the condensations move outward: Comparison of the plasma velocity in the compression region with the expansion velocity of the outer boundary, $R_1(t)$, shows that the escape of plasma along the dipole axis from the magnetic-compression region can be neglected up to time τ_n , at which the magnetic forces become ineffective. Accordingly, at time τ_n , at which the leading edge of the envelope has overtaken the outer boundary of the raking region, the polar condensations are compact plasma formations at a distance R_{1max} from the center of the envelope and moving outward in opposite directions at ~ 150 km/sec. Using the assumptions above we find $R_{1max} \sim 10^4 R_0$, so that with $R_0 \sim R_\cdot \sim 7 \cdot 10^{10}$ cm we would have $R_{1max} \sim 7 \cdot 10^{14}$ cm. Nearly all the gas from the raking region is accumulated in the polar condensations, so that the mass of these condensations should be $M_k \sim 5 \cdot 10^{21}$ g.

The time over which the raking region exists is the time over which the polar condensations are formed, equal to $\tau_k \sim 2 \cdot 10^7$ sec under our assumptions.

These results agree with the kinematics of the polar condensations in the envelope of DQ Her 1934. However, the mass found for these condensations is too small a fraction of the mass of the envelope, $M_{en} \sim 10^{29}$ g [30]. The reason for this contradiction apparently lies in the influence of the magnetic field on the leading part of the envelope. In this connection it is interesting to consider another property of the envelope.

We assume that the condition $H^2/8\pi = \rho V^2/2$ is satisfied at some time t on some Lagrange sphere M within the expanding envelope. Then at a later time $t + \Delta t$, on the same Lagrange sphere M, we would have $H^2/8\pi < \rho V^2/2$, since we have $H^2/8\pi \sim r^{-4}$ but $\rho V^2/2 \sim r^{-3}$. At the earlier time $t - \Delta t$ the same Lagrange sphere M was under the control of magnetic forces: $H^2/8\pi < \rho V^2/2$. Because of this property we can treat the outer edge of the envelope at large values of t as the inner boundary of the magnetic-raking region. On the other hand, at small values of t this property shows that the assumption made regarding the initial radius of the envelope – it was assumed equal to the radius of the star at the time of the explosion – is important. If the explosion occurs from a state which is denser (which has a smaller value of R_0) than assumed above, a significant part of the envelope can come under the control of magnetic forces. However, study of this question requires additional assumptions regarding the structure of the inner part of the envelope and, ultimately, regarding the mechanism for its ejection.

Our purpose here has been to study the raking of the interstellar plasma by the magnetic field of the expanding envelope, and we have reached the following conclusion: This raking leads to the appearance of condensations of the interstellar plasma ahead of the envelope. The expanding envelope subsequently overtakes these condensations, but since their mass is small their influence on the envelope structure can be neglected. The envelope structure must be explained on the basis of other arguments, e.g., an influence of the magnetic field on the expansion of the envelope itself.

§ 4. Magnetic Raking of Plasma as a Possible Mechanism for the Formation of Certain Types of Solar Prominences

Raking of plasma by a strong dipole magnetic field can probably explain the formation of certain types of chromospheric outbursts (see the next subsection) and the initial stage in the formation of coronal condensations [54]. Phenomena analogous to solar coronal condensations should occur at other stars, especially magnetic variable stars. Indirect evidence in favor of this assertion comes from the observation that for certain magnetic variable stars the maxima of the magnetic field coincide with brightness maxima [55-57].

A. Observations and Classification of Solar Prominences

Among the wide variety of explosive events in the solar atmosphere which are accompanied by plasma ejection above active regions, we can distinguish two types of prominences [58, 59]: Prominences of the first type ("sprays") usually accompany flares and are characterized by the expansion of the ejected plasma from the site of the flare over a broad solid angle at a high velocity (~1000 km/sec). The prominences of the second type ("surges") [60, 61], on the other hand, are quasiperiodic, have a narrower directional pattern, and have a lower characteristic velocity (~100 km/sec).

If the sprays are consequences of flares, there is apparently no direct relationship between surges and flares, although both phenomena are undoubtedly due to rapidly changing magnetic fields, so that they frequently occur at the same times and in the same regions, above active regions. However, there have been observations of surges in the absence of noticeable flares [62].

Flares and the accompanying sprays are known to be characterized by bipolar magnetic fields. Surges usually appear at the outer boundary of the penumbra of young spots [61, 63] or even outside the spots, above the nucleating pores [64]. There is no direct evidence linking surges with sign-changing magnetic fields. Furthermore, Fortini and Torelly [62] have reported observation of a large number of surges near a rapidly growing spot within a unipolar group.

Below we will be concerned with surges only. Typical properties of surges are as follows: an average density of $\sim 10^{12}$ cm^{-3}, a height of $\sim 10^{9}$ cm, a thickness of $\sim 5''$ ($\sim 3 \cdot 10^{8}$ cm), a mass of $\sim 10^{14}$ g, a surge time of ~5-10 min, and a velocity of $\sim 10^{7}$ cm/sec.

B. Theoretical Models for Surges

Several investigators (see, e.g., [63, 65]) have attempted to explain surges by exploiting an analogy with flares and sprays: The reconnection of magnetic fields of opposite polarity is accompanied by the heating of plasma, which is explosively ejected along magnetic lines of force. This argument was criticized in [66]; in the present discussion we will cite only some of the objections against the magnetic-reconnection model: First, observations [67] show that the temperature in a surge is not at the high level required according to this model; second, magnetic reconnection tends to destroy the original magnetic field configuration, while observations show that surges frequently occur repeatedly in the same region [68]; finally, since surges are a relatively frequent phenomenon, it seems unlikely that they would have to be preceded by a complicated and very specific magnetic field configuration.

Schlüter [69] proposed a mechanism for surges which does not require magnetic reconnetion. According to this proposal, a magnetic field of a divergent configuration (a tube of lines of force heading upward from the photosphere) ejects a blob of diamagnetic plasma. It should be kept in mind, however, that this process does not explain the opposite (downward) plasma motion and thus has no bearing on the actual situation at the sun.

Altschuler et al. [66] attempted to circumvent this difficulty of the preceding model. They assume that there is a current ring within the plasma blob which produces an internal magnetic field (in the preceding model the surface currents prevented the field from penetrating into the blob), and they solve the MHD problem of the ejection of this ring. Now there is a magnetic field everywhere, and some of the plasma is in fact ejected (along with the current ring) upward at the Alfvén velocity. However, most of the mass of the blob goes downward, leading to quantitative difficulties in the explanation for the surges, not to mention the circumstance that the model does not justify the initial state from which the surge began.

In all these models the surges are linked to an immediate lifting of material into the corona. However, there is another possible way to explain surge formation: It could be argued that the surges appear as a result of the raking of coronal matter accompanying an intensification of a local magnetic field [70, 71]. It is a simple matter to give a qualitative explanation for this process: Some dipole moment can be associated with any diverging tube of magnetic lines of force. If this dipole moment is increasing (if the field is increasing or if new lines of force are being added), the plasma is plowed toward the axis of the dipole (or tube) and is simultaneously accelerated along lines of force.

The surge probably occurs in two stages [70]: In the first stage the plasma is raked and accelerated rapidly (over ~1 min). The gravitational force is inconsequential in this stage. In the second stage the plasma continues its inertial motion upward along lines of force until it is stopped in the gravitational field (~5-10 min), at which time it begins to fall again. Let us discuss the first stage in more detail.

C. Magentic Raking as a Mechanism for Surges

Let us consider the process leading to the appearance of surges at the outer boundary of the penumbra of growing spots. The growth of spots is usually accompanied by an increase in their area, while the magnetic field within a spot remains relatively constant. It is thus natural to assume that the appearance of surges results from the attachment of new lines of force to the preexisting magnetic field of the spot. We assume that a pencil of new lines of force is attached to the magnetic field of the spot at some part of the outer boundary of the penumbra. The local motion of these lines of force is of the same nature as the motion of the lines of force of an increasing magnetic dipole. Generally speaking, the axis of such a dipole should be inclined at some angle α with respect to the axis of the spot, and this angle may vary with the time. Ignoring this possible time dependence for the moment, we assume that the dipole is perpendicular to the photosphere (an individual growing pore would probably be the ideal example). What kind of density profile can an increasing magnetic dipole of this type produce above itself?

We assume that the plasma density profile above the photosphere is initially exponential:

$$\rho_0(h_0) = \rho_{00} \exp(-h_0/h_{00}). \tag{1.83}$$

Here ρ_{00} is the density at some level in the photosphere, which we assume to be locally planar, h_0 is the height above the photosphere, and h_{00} is the scale height ($h_{00} = kT/mg \sim 3 \cdot 10^7$ cm for T ~ 6000°K and g ~ $3 \cdot 10^4$ cm/sec^2).

How does this density profile change upon an increase in the moment of a dipole whose axis is perpendicular to the photosphere? As the dipole moment increases by a factor of m/m_0, the constant-density surface is described by

$$\rho(r, \theta, \varphi, m) = \rho_0(r_0 \cos\theta_0, \varphi_0)\,\mathscr{P}(m, \theta) \equiv \tilde{\rho} = \text{const}, \tag{1.84}$$

where

$$r_0 = r_0(r, \theta) = \frac{r}{\mathscr{R}(m, \theta)}, \qquad \theta_0 = \theta_0(m, \theta), \quad \varphi_0 = \varphi \tag{1.85}$$

are the initial coordinates of the fluid particle; and r, θ, φ are the coordinates of the same particle at time t, when the dipole moment is m. The functions $\mathscr{P}(m, \theta)$, $\mathscr{R}(m, \theta)$, and $\theta = \theta(m, \theta_0)$ are defined by (1.71), (1.69), and (1.67), respectively.

For the exponential density profile in (1.83) we find from (1.84) the equation of the family of constant-density surfaces:

$$r(\theta, m) = h_{00} \frac{\mathscr{R}(m, \theta)}{\cos\theta_0(m, \theta)} \left[\ln \mathscr{P}(m, \theta) - \ln \frac{\tilde{\rho}}{\rho_{00}}\right]. \tag{1.86}$$

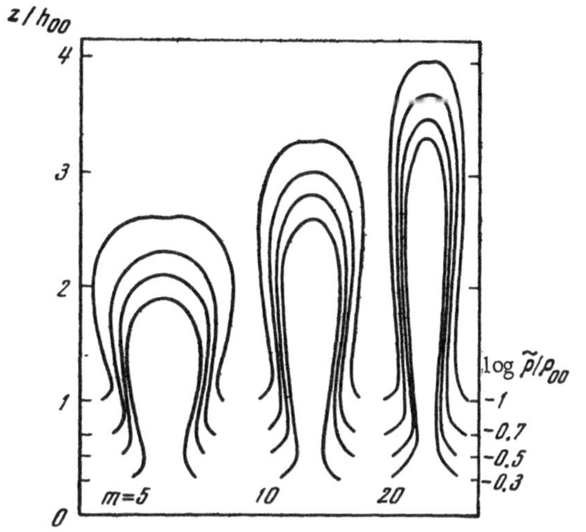

Fig. 3. Density profiles of the plasma raked by a vertical growing magnetic dipole in an expotential atmosphere.

Figure 3 shows the meridional cross sections of these surfaces for several values of the dipole moment m (m = 5, 10, 20).

For the case of a sloping dipole the surge evolves in a slightly different manner. If the dipole axis makes an angle α with the vertical, and the magnetic moment is increased by a factor of m/m_0, the equation of the family of constant-density curves in the vertical plane passing through the dipole axis becomes

$$r(\theta,\ m)=h_{00}\ \frac{\mathscr{R}(m,\ \theta)}{\cos\left[\theta_0(m,\ \theta)+\alpha\right]}\left[\ln\mathscr{P}(m,\ \theta)-\ln\frac{\tilde{\rho}}{\rho_{00}}\right]. \tag{1.87}$$

Figure 4 shows the pattern of these curves for α = 30°; m = 5, 10, 20; and $\ln(\tilde{\rho}/\rho_{00}) = 0$. We see that the change in the density in the case of a sloping dipole is more complicated, accompanied by a "straightening" of the surge above the photosphere.

For comparison, Fig. 5 shows several successive contours of a growing surge observed on 23 X 1970 [71].

Interestingly, most of the surges have a slight central dip at the maximum phase of the evolution. To explain this dip we turn to the velocities of fluid particles at the "crest" of the

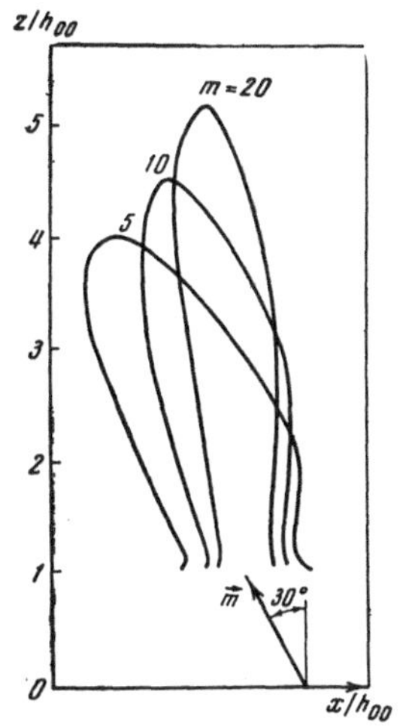

Fig. 4. Density profiles of a plasma raked by a sloping (α = 30°) growing magnetic dipole in an exponential atmosphere.

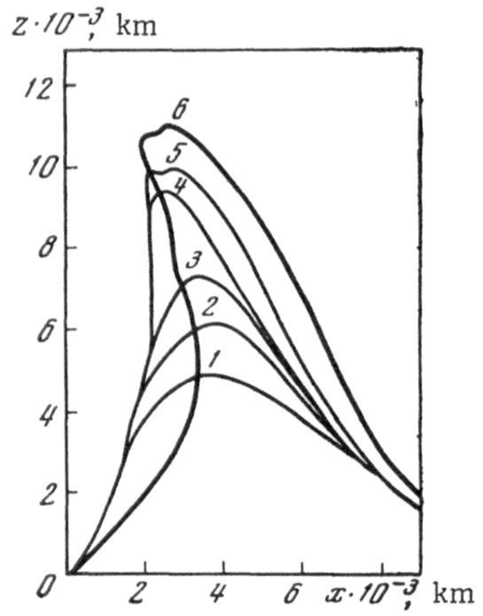

Fig. 5. Observed constant-density contours at successive times (1-6) in a surge according to the data of [71].

surge at the instant the magnetic field stops increasing. According to (1.68), the radial component of the dimensionless velocity of a fluid particle is

$$v_r(r,\ \theta,\ t)=\frac{r(\theta,\ m)}{\mathscr{R}(m,\ \theta)}\dot{\mathscr{R}}(t,\ \theta)=\frac{r(\theta,\ m)}{\mathscr{R}(m,\ \theta)}\frac{d\mathscr{R}}{dm}\dot{m}. \tag{1.88}$$

The derivative $\dot{m}$ can be estimated roughly on the basis of the velocity of the outer density contour, i.e., of the boundary of the surge, which we define by the condition $\tilde{\rho}=\rho_{00}$. Then setting $\theta_0=0$ in (1.86) and assuming $\mathscr{P}(m,\ 0)=m/m_0$, $\mathscr{R}(m,\ 0)=1$, at the dipole axis, we find

$$r(0,\ m)=h_{00}\ln\frac{m}{m_0}. \tag{1.89}$$

Hence

$$\dot{m}=\frac{m}{h_{00}}\dot{r}(0,\ m), \tag{1.90}$$

and the value of m can be evaluated from the observed height of the surge, (1.89), at the end of the first (acceleration) stage ($\sim t_1\sim 1$ min).

From the observational data [70] we find $r(0,\ m)\sim 10^8$ cm and $\dot{r}(0,\ m)\sim 10^7$ cm/sec, so that we have $m/m_0\sim 10$; Fig. 6 shows the corresponding profile of the radial velocity component in the upper part of the surge at the time the magnetic field stops increasing.

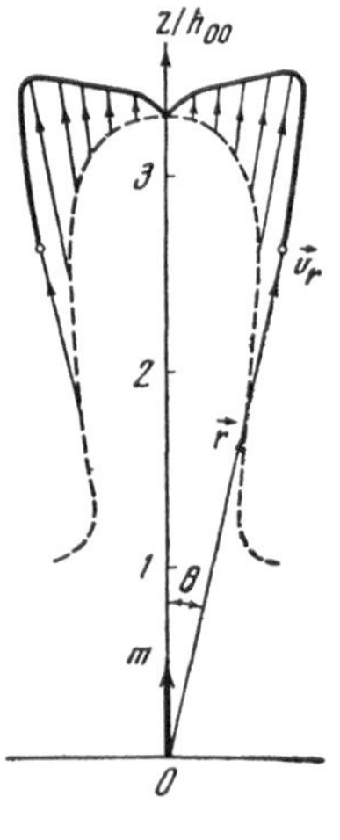

Fig. 6. Density profile for $\bar{\rho}/\rho_{00}=1$ with m = 10 (dashed curve) and corresponding profile of the dimensionless radial velocity v_r as functions of the polar angle θ.

The subsequent inertial motion of the plasma along lines of force (the second stage) produces the dip at the crest of the surge.

Evidence in favor of this model comes from spectral observations [70] which show that the transverse velocity component v_θ corresponds to raking of the plasma toward the axis of the surge.

Since this discussion is based on the assumption that the approximation of a strong field and a cold plasma is valid for surges, i.e., on the assumption of conditions (19), we should take a look at the validity of these assumptions. Adopting V ~ 40 km/sec as a typical flow velocity of the plasma [70], we find $V^2 \ll V_A^2$ ($\varepsilon^2 \ll 1$) for a magnetic field of H ≫ 20 G and for a density of matter $\rho \sim 10^{-12}$ g/cm^3. The second condition, $V_s^2 \ll V^2$ ($\gamma^2 \ll \varepsilon^2$), holds if the plasma temperature is $T \ll 2 \cdot 10^5$ °K.

To obtain the observed surge velocity we require rapid changes in the magnetic moment (over times on the order of 10 sec); unfortunately, we do not have any data at present which would permit us to reach a conclusion regarding this point.

This proposed mechanism for the raking of the plasma in the solar atmosphere upon a local intensification of the photospheric magnetic field probably operates in a broad class of solar phenomena, ranging from small-scale phenomena such as spicules to large-scale phenomena such as polar condensations. In the latter case the reason for the raking of the plasma should be identified as a change in the magnetic field in a large unipolar region [54].

CHAPTER II

APPEARANCE OF A (NEUTRAL) CURRENT SHEET DURING PLASMA MOTION IN THE FIELD OF A PLANE MAGNETIC DIPOLE

§ 1. Condition for the Absence of Continuous Two-Dimensional Plasma Flows with a Strong Frozen-in Magnetic Field; Appearance of a Current Sheet

The significance of magnetic null points and of the current sheets which develop from them has been discussed in connection with the problem of magnetic energy dissipation and particle acceleration in a plasma (see [16, 17, 19, 72-75] and the literature cited there). Syrovatskii [19] analyzed the conditions for the appearance of (neutral) current sheets in two-dimensional flow of plasma with a strong, frozen-in magnetic field, finding that the presence of singular magnetic null points (H = 0, E ≠ 0) precludes continuous plasma flow near these points because the frozen-in condition is violated there (see Chapter I, §1). As a result, current sheets appear at those places in a plasma in a strong magnetic field at which singular null points (lines parallel to the z axis) should appear in the absence of a plasma.

Syrovatskii [19] pointed out a method for constructing a current sheet and the surrounding magnetic field: A current sheet must be placed at the singular null line in such a manner that it passes through all the secondary null lines which arise if a line current is placed at the original null line and if this line current varies from zero to some finite value; this value determines the width of the current sheet (Fig. 7). Then a potential magnetic field [an analytic function (1.14) on the complex plane] is sought which satisfies the given boundary conditions at the boundary of the region and which has a vanishing normal component at the surface of the current sheet (at the cut in the complex plane).

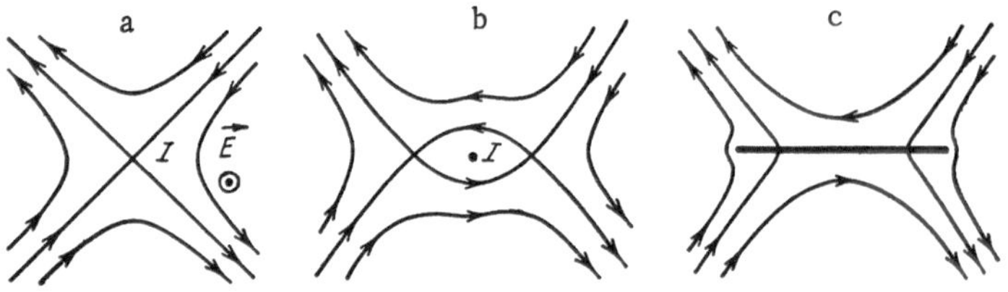

Fig. 7. Formation of a neutral current sheet at a singular magnetic null line. a) Magnetic lines of force near the original null line and direction of the external electric field; b) pattern of magnetic lines of force corresponding to the "splitting" of the null line as a line current passes along it in the case in which the magnetic field is not frozen in the plasma; c) current sheet formed in the interaction of a line current with an external magnetic field which is well frozen in the plasma.

Below we apply this method to the case in which the current sheet arises as a result of the capture and stretching of a dipole magnetic field by a plasma flow.

§ 2. Problem of the Appearance of a Current Sheet as a Plasma Moves in the Field of a Plane Magnetic Dipole

A. Possible Applications of the Problem

We can cite three possible applications of the problem. One would be the tail of the earth's magnetosphere. In the tail there is known to be a so-called neutral sheet separating magnetic fields which are opposite in direction [76] (see also the reviews in [20, 21, 77-79]). For a strong magnetic field and two-dimensional geometry the concepts of a neutral sheet and a current sheet are identical, since if the plasma pressure is negligibly low any current sheet must separate regions in which the magnetic fields are equal in magnitude but opposite in direction. Accordingly, we are equally justified in using both terms below.

The reasons for the appearance of this magnetic tail have not been finally resolved [20, 21]; we will see below that even a partial penetration of the geomagnetic field into the plasma of the solar wind turns out to be sufficient for the appearance of a current sheet.

The second field of application of these results may be coronal rays or streamers. Observations of the solar corona during eclipses reveal that the coronal matter is typically distributed in coronal rays or streamers and helmet-shaped formations or plumes [80, 81]. It is now clear that these formations are related to the local large-scale magnetic field at the solar surface; this relationship has been verified by direct calculations of the magnetic field in the corona based on data on the radial component of the field in the photosphere [82, 83]. These calculations are based on two main assumptions: Above the photosphere the magnetic field is a potential field up to some level in the corona, at which, as a result of the entrainment of the magnetic field by the outward moving solar wind, the field becomes purely radial. The magnetic field distributions calculated on the basis of these assumptions correlate well with the optical structure of the chromosphere and corona, radio and x-ray images of the sun [84, 85], and the structure of the interplanetary field [86].

However, this correlation holds for only the general field configuration. It is not difficult to see that the magnetic field constructed by the methods developed in [82, 83] should in general contain null points; this circumstance was not taken up in [82, 83]. As Syrovatskii has shown [19], the appearance of null points corresponds to a special situation in which the presence of the plasma cannot be neglected, even if its density and pressure are negligibly low. It is precisely because of the high plasma conductivity near null points that current sheets with the corresponding magnetic-field geometry should arise.

A similar situation – involving a current sheet flanked by quasiradial fields in opposite directions – is apparently observed in streamers. Formally, the appearance of coronal streamers is completely analogous to the appearance of the magnetospheric tail. In both cases a dipole magnetic field is stretched out by the plasma flow of the solar wind; in the corona it is the dipole magnetic field of an extended active region, while in the magnetosphere it is the geomagnetic field.

We should immediately stipulate that the model developed below is useful only as a first extremely crude approximation of the actual situation, primarily because of the assumptions that the situation is two-dimensional and that the field is strong. We are obviously stretching the facts a bit when we classify the tail of the magnetosphere or streamers as plane, two-dimensional formations. Furthermore, far from the dipole the field weakens, the solar-wind energy becomes comparable to the magnetic energy, and the field can no longer be assumed a potential field.

Nevertheless, despite these limitations, the proposed method demonstrates the physical nature of the mechanism for the appearance of formations such as streamers and the geomagnetic tail, and in certain cases it is even useful for quantitative calculations. This brings up the third possible application of the results: laboratory experiments involving plasma flow around a plane magnetic dipole [24-26]. The purpose of these experiments has been to simulate the conditions in the magnetosphere in a two-dimensional geometry. In this latter regard they best resemble the following formulation of the problem, with the limitation that under laboratory conditions the need to ensure a sufficient freezing of the magnetic field in the plasma poses a serious problem.

B. Formulation of the Problem. Field in the Absence of a Plasma

We use the MHD approximation of a strong field, assuming conditions (14), i.e.,

$$V_s^2 \ll V_A^2, \qquad V^2 \ll V_A^2, \tag{2.1}$$

where V_s is the gasdynamic sound velocity, V is the typical value of the plasma flow velocity, and V_A is the Alfvén velocity.

According to [19], for two-dimensional problems the magnetic field should be a potential field throughout the region in which conditions (2.1) hold, except possibly at isolated surfaces (cuts in the complex plane), i.e., at current sheets.

As a model for the stretching of a field by the solar wind we consider the following idealized problem [87]: We assume a two-dimensional magnetic dipole at the base of a semicylindrical region on a complex plane (Fig. 8). We assume that conditions (2.1) hold within this region and that magnetic flux is conserved at the boundary R of the region, which is expanding in accordance with a given law R = R(t). We assume that the dipole field partially penetrates this boundary, so that the initial magnetic flux at each surface point is a fraction α of the flux in the absence of a boundary.

For the case of the solar corona this boundary represents the transition region from the chromosphere and the lower corona, where the strong-field conditions hold, to the upper

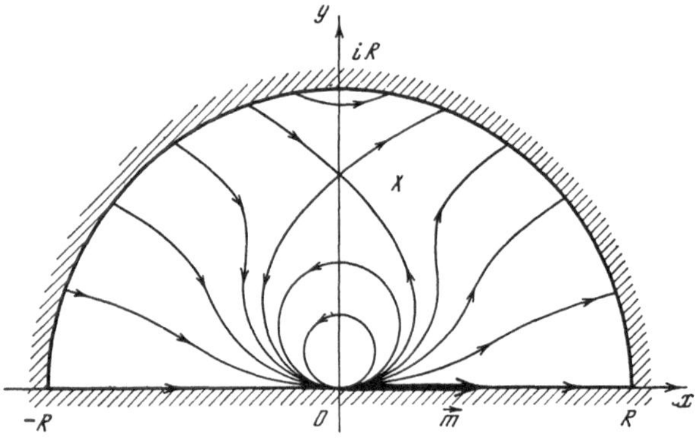

Fig. 8. Magnetic field of a two-dimensional dipole with lines of force frozen in the plasma at the boundary of a semicylindrical region. X) Null point (line along the z axis) of the magnetic field.

corona, where the energy of the solar-wind plasma dominates. The magnetic field of the active region generating the helmet-shaped formations and coronal rays is approximated by the field of a two-dimensional magnetic dipole.

For a magnetosphere with a steady-state magnetic tail this problem can serve as a model for the appearance of a tail on an originally spherical or quasispherical magnetosphere.

In this formulation of the problem the vector potential is governed by the Laplace equation

$$\Delta A = 0 \tag{2.2}$$

with the boundary conditions

$$A(x, y, t) \equiv A(r, \varphi, t) = \begin{cases} 0 & \text{at} \quad y = 0, \quad -R \leqslant x \leqslant R, \\ \dfrac{\alpha m}{R_0} \sin \varphi & \text{at} \quad r = R, \quad 0 \leqslant \varphi \leqslant \pi \end{cases} \tag{2.3}$$

and a dipole singularity at the origin,

$$A(r, \varphi, t) \to \frac{m \sin \varphi}{r} \quad \text{as} \quad r \to 0. \tag{2.4}$$

Here r and φ are the polar coordinates in the (x, y) plane, R_0 is the initial value of R(t), α is the fraction of the magnetic flux of the dipole which penetrates the boundary, and m is the magnitude of the dipole magnetic moment, which we assume below to be constant.

Since the magnetic flux through contour element dl = {dx, dy} is dA [see (1.8)], condition (2.3) is a consequence of the symmetry of the problem; thus the flux across the x axis vanishes, while at the surface r = R(t) condition (2.3) corresponds to conservation of the initial (at r = R_0) flux through each element of the expanding cylindrical surface.

The solution of problem (2. 2)-(2. 4) in terms of the complex potential in (1. 14) is obvious:

$$F(z, t) = \frac{im}{z} - \frac{im(\alpha R - R_0)}{R^2 R_0} z. \tag{2.5}$$

The first term describes the field of a plane dipole, while the second corresponds to that effective uniform field which is necessary to satisfy boundary condition (2. 3).

If, as it increases, the radius R(t) of the region becomes larger than $2R_0/\alpha$, then a type X magnetic null point appears on the y axis (Fig. 8), at

$$z_0 = iR\left(\frac{R_0}{\alpha R - R_0}\right)^{1/2}. \tag{2.6}$$

The electric field at the null point does not vanish:

$$F_z = \frac{m}{cR_0}\frac{\alpha R - 2R_0}{R^2(\alpha R/R_0 - 1)} R > 0 \tag{2.7}$$

for $\dot{R} > 0$ and $R > 2R_0/\alpha$. Accordingly, the null point of (2.6) is a singular point and must be eliminated from the region in which solution F(z, t) is defined, by introducing a cut in the complex plane corresponding to the current sheet.

C. Solution with a Current Sheet

In accordance with the rule found in [19] and outlined above, a singular null point which appears in the case $R > 2R_0/\alpha$ must be eliminated from the region in which the solution is defined by introducing a cut Σ in the complex plane. This cut should be made along the y axis, from the upper boundary of the region in which the null point first appears, to some height h (Fig. 9a). The length of the cut at a given time (i.e., the distance $R - h$) must be no shorter than the minimum length required to eliminate the null point.

If there is a current sheet, a boundary condition must be specified at the edges of the cut. We assume that the lines of force do not intersect the cut, i.e., that

$$A|_\Sigma = A_0(t). \tag{2.8}$$

Neglecting possible dissipative processes at the current sheet, we assume $A|_\Sigma$ = const, specifically [see (2.3)]

$$A|_\Sigma = \alpha m/R_0. \tag{2.9}$$

It is therefore necessary to solve the Dirichlet problem for Eq. (2.2) with conditions (2.4) and

$$A(x, y, t) = \begin{cases} 0 & \text{at} \quad y = 0, \quad -R \leqslant x \leqslant R; \\ (\alpha m/R_0)\sin\varphi & \text{at} \quad r = R, \quad 0 \leqslant \varphi \leqslant \pi; \\ \alpha m/R_0 & \text{at} \quad x = 0, \quad h \leqslant y \leqslant R. \end{cases} \tag{2.10}$$

To solve this problem we use a symmetry principle (see p. 143 in [27]); specifically, we supplement the region in which the solution is defined with the region which is symmetric

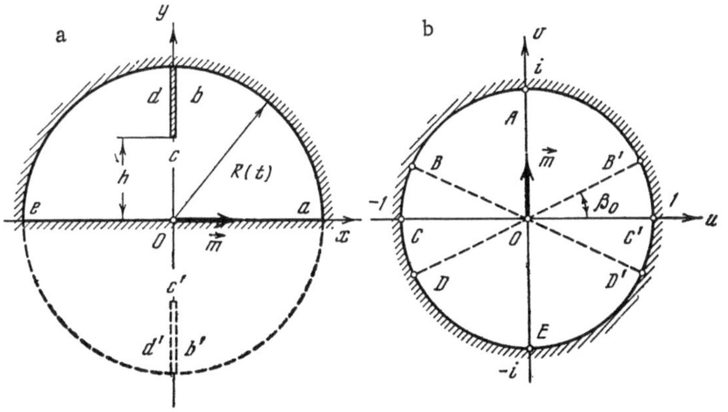

Fig. 9. a) Elimination of the singular null point by means of a cut in the complex plane and a symmetric expansion of the range of applicability of the solution. The hatching shows the position of the new boundary; b) mapping of the symmetrized region into the interior of a unit circle.

with this first region with respect to the x axis (the dashed lines in Fig. 9a). Correspondingly, we replace boundary condition (2.10) by

$$A(x, y, t) = \begin{cases} am/R_0 & \text{at} \quad \Sigma, \\ (am/R_0)\sin\varphi & \text{at} \quad r = R, \\ -am/R_0 & \text{at} \quad \Sigma'. \end{cases} \tag{2.11}$$

We seek a solution in the form

$$F(z, t) = \frac{im}{z} + f(z, t), \tag{2.12}$$

where

$$f(z, t) = a(x, y, t) + ib(x, y, t) \tag{2.13}$$

is an unknown analytic function. By virtue of (2.11) the real part of this function must satisfy the boundary condition

$$a(x, y, t) = \begin{cases} am/R_0 - m/y & \text{at} \quad \Sigma, \\ (am/R_0 - mR)\sin\varphi & \text{at} \quad r = R, \\ -am/R_0 - m/y & \text{at} \quad \Sigma'. \end{cases} \tag{2.14}$$

Using the function (see [27])

$$w(z, t) = u + iv = \frac{i\cos\beta_0}{2}\left\{\left(\frac{z}{R} - \frac{R}{z}\right) - \left[\left(\frac{z}{R} - \frac{R}{z}\right)^2 + \frac{4}{\cos^2\beta_0}\right]^{1/2}\right\}, \tag{2.15}$$

where

$$\cos\beta_0 = \frac{2\varkappa}{1+\varkappa^2}, \qquad \varkappa = \frac{h}{R}, \qquad 0 \leqslant \beta_0 \leqslant \frac{\pi}{2}, \tag{2.16}$$

we conformally map the region with cuts (Fig. 9a) into the interior of a unit circle (Fig. 9b). The sign in front of the square bracket in (2.15) (a minus sign when the root is real and positive) determines which branch of the Zhukovskii function we must choose for the mapping into the interior of a unit circle. We state for clarity that with $\beta_0 = 0$ we have

$$w(z, t) = \frac{iz}{R}. \tag{2.17}$$

Using (2.15) to transform the boundary conditions in (2.14), and solving the problem in the w plane, we find the general solution [87]

$$F(w(z, t), t) = f_0(w, t) + f_1(w, t) + f_2(w, t), \tag{2.18}$$

where

$$f_0(w, t) = \frac{im}{z} = -\frac{2m\cos\beta_0}{R}\left\{\left(w + \frac{1}{w}\right) - \left[\left(w + \frac{1}{w}\right)^2 - 4\cos^2\beta_0\right]^{1/2}\right\}^{-1}, \tag{2.19}$$

$$f_1(w, t) = \frac{2m}{R\cos\beta_0}w - \frac{m}{2R\cos\beta_0}\left\{\left(w + \frac{1}{w}\right) - \left[\left(w + \frac{1}{w}\right)^2 - 4\cos^2\beta_0\right]^{1/2}\right\}, \tag{2.20}$$

$$f_2(w, t) = -\frac{am}{R_0\cos\beta_0}w - \frac{am}{R_0\cos\beta_0}\frac{2}{\pi}\left\{\frac{i}{2}\cos\beta_0 \ln\frac{1 - w^2 - 2iw\sin\beta_0}{1 - w^2 + 2iw\sin\beta_0} + \right.$$
$$\left. + \frac{1}{2}\left(w + \frac{1}{w}\right)\frac{i}{2}\ln\left[(1 - w^2e^{-2i\beta_0})/(1 - w^2e^{2i\beta_0})\right] - \beta_0 w\right\}. \tag{2.21}$$

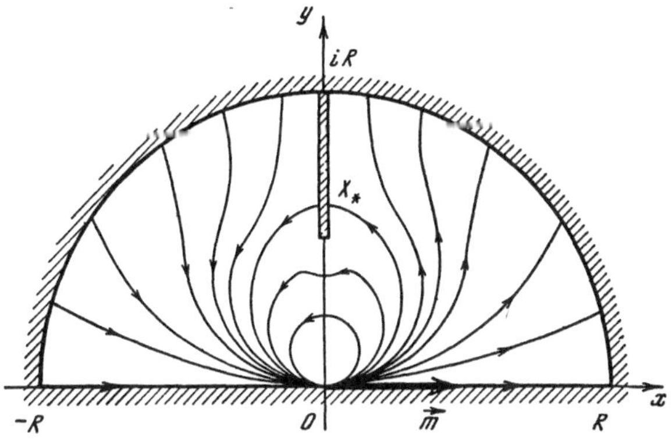

Fig. 10. Pattern of magnetic lines of force corresponding to the solution of the problem with a current sheet.

Figure 10 shows schematically the pattern of magnetic lines of force corresponding to this solution. The magnetic field vanishes at the surface of the current sheet at point X_* with coordinates $x = 0$, $y = h_*$, where h_* is defined by

$$\frac{\partial F}{\partial w}(e^{i\beta_*}) = 0. \tag{2.22}$$

The null point X_* is a nonanalytic singular point lying at the boundary of the region of applicability for this solution. The potential near this singular point is

$$F(z, t) \simeq \pm \frac{1}{2} C_1 (z - ih_*)^2 e^{-i\pi/4} + C_2, \qquad x = \pm 0, \tag{2.23}$$

where C_1 and C_2 are (generally time-dependent) constants and $x = +(-)0$ corresponds to the right-hand (left-hand) edge of the cut.

The null point X_* separates a region of "forward current" ($x = 0$, $h_* \leq y \leq R$) from a region of "return current" ($x = 0$, $h \leq y < h_*$). In general the solution is bracketed by two limiting regions: (1) that in which the return current is zero; (2) that in which the total current in the cut is zero (the return current cancels the forward current).

The first case corresponds to the minimum length of the current sheet, i.e., to the largest height h. To find this height we must solve the equation

$$\frac{\partial F}{\partial w}(-1) = 0 \tag{2.24}$$

for β_0. The solution shows that β_0 depends on only the combination $\alpha R/R_0$ and is

$$\beta_{01} = \frac{\pi}{2}\left(1 - \frac{2R_0}{\alpha R}\right). \tag{2.25}$$

Here the subscript "1" shows that the equation corresponds to the first case. From Eqs. (2.16) and (2. 25) we find

$$h_1 = R \tan \frac{\pi R_0}{2\alpha R}. \tag{2.36}$$

Figure 11 shows the corresponding pattern of lines of force. The magnetic field vanishes at the lower end of the current sheet, at point Y. Near this point we have

$$F(z, t) \simeq d_1 (z - ih)^{3/2} + d_2, \tag{2.27}$$

where d_1 and d_2 are time-dependent constants.

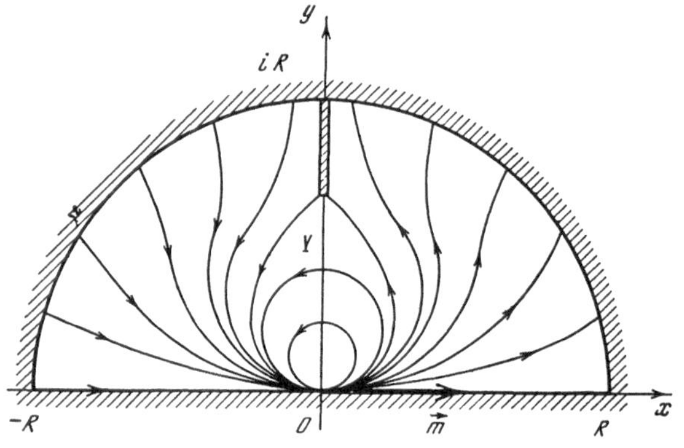

Fig. 11. Pattern of magnetic lines of force in the particular case in which there is no return current (the length of the cut is minimal).

Let us consider a second limiting case. A zero total current implies a zero circulation over the contour Γ around the cut Σ. The result is a transcendental equation for β_{02}:

$$\beta_{02}+\frac{1}{2}\cot\beta_{02}\ln\cos\beta_{02}=\beta_{01}. \tag{2.28}$$

At small values of β_{02} (i.e., for short cuts or for R differing only slightly from $2R_0/\alpha$), in the linear approximation, we can write

$$\beta_{02}\simeq\frac{4}{3}\beta_{01}. \tag{2.29}$$

The pattern of force lines is shown schematically in Fig. 10.

As R increases, the solution rapidly assumes its asymptotic form; to find this asymptotic form we must take the limit $\alpha R/R_0 \to \infty$ in (2.18). We find

$$F(w, t)=\frac{m}{2h}\left(w-\frac{1}{w}\right)-\frac{2im\alpha}{\pi R_0}\ln\frac{1-iw}{1+iw}, \tag{2.30}$$

$$w(z, t)=-i\left[\frac{h}{z}+\left(\frac{h^2}{z^2}+1\right)^{1/2}\right], \tag{2.31}$$

$$\cos\beta_0=0, \qquad \beta_0=\frac{\pi}{2}. \tag{2.32}$$

For the case in which the return current vanishes, we find from (2.25) and (2.26)

$$\beta_{01}=\frac{\pi}{2}, \qquad h_1=\frac{\pi}{2\alpha}R_0. \tag{2.33}$$

In this second limiting case, with a vanishing total current, β_{02} is thus given by

$$\beta_{02}+\frac{1}{2}\cot\beta_{02}\ln\cos\beta_{02}=\frac{\pi}{2}, \tag{2.34}$$

from which we find

$$\beta_{02}=\frac{\pi}{2}, \qquad h_2=0. \tag{2.35}$$

Figure 12 shows the patterns of lines of force in the case $\alpha = 1$ for $h = 0.8R_0$ and $h = h_1 = \pi R_0/2$. Near the dipole the magnetic field has the same structure as in the general case, but far from the dipole the lines of force tend toward radial straight lines.

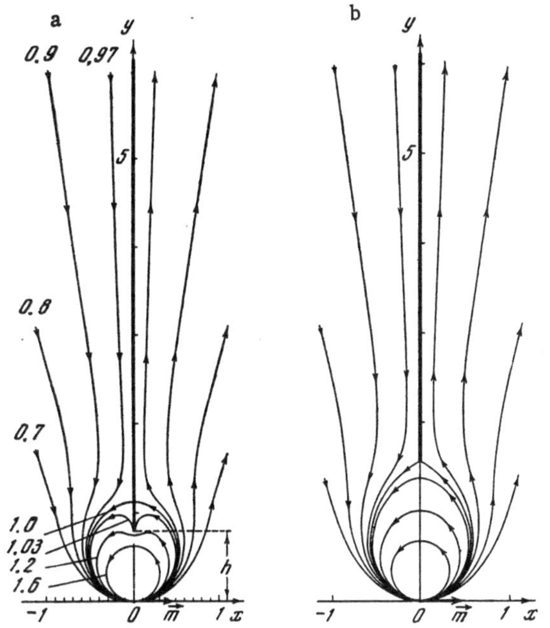

Fig. 12. Magnetic lines of force. a) Lines corresponding to the asymptotic solution for $\alpha = 1$, $h = 0.8$ (R_0 is adopted as the unit of length). The curves are labeled with the value of the potential A; the unit here is m/R_0. Below the point X_* on the cut is the region of return currents which cause the characteristic sag in the magnetic lines of force; b) for the particular case in which there is no return current, and the length of the cut is minimal ($h = h_1 = \pi/2$).

D. Discussion

One advantage of this model is that it permits complete calculations without loss of the basic physical content of the process. As a result, the conditions under which neutral sheets appear during plasma flow in a dipole magnetic field become clearer. Three of these basic conditions are as follows:

(1) The plasma conductivity must be high enough that we can assume the magnetic field to be frozen in the plasma.

(2) There must exist a boundary or transition layer between the region in which magnetic tension is governing (the magnetic cavity) and the region where the plasma energy is governing (the solar wind).

(3) The magnetic field must penetrate from the magnetic cavity into the wind; i.e., the field must be "trapped" by the wind.

The first two of these conditions are obvious, but a few comments are in order regarding the third.

In the case of the solar corona the field is captured by the solar wind "inside" the field itself: In the strong-field region the matter is flowing slowly along lines of force; then, as the field becomes weaker, the flow converts into the radial solar wind which carries off the outer part of the field with itself. As a result, for a long-lived active region a quasisteady pattern of the outward flow of material along coronal rays is established [88, 89]. A rigorous calculation of the steady-state regime must be based on a consistent analysis of the magnetic field and the solar wind.

In the case of the magnetosphere the trapping of the magnetic field by the solar wind is probably due to an instability of the interface between the wind and the magnetosphere. Such an instability could be caused, for example, by the instability of a tangential velocity discontinuity [2, 90].

As is clear from the discussion above, a current sheet arises if there is an arbitrarily slight penetration of the field into the wind, i.e., for arbitrarily small values of α in Eq. (2.30).

For quasisteady phenomena such as the magnetospheric tail and coronal rays, the situation is evidently close to the limiting case of no return current, since this return current manages to decay in the course of the slow growth of the sheet [19]. In this case, according to (2.33), as α decreases the beginning of the sheet moves away from the magnetic dipole, and the flow toward the tail decreases. Accordingly, the effectiveness of the penetration of the magnetospheric field into the solar wind must be high.

If the existence of a neutral sheet for the magnetosphere has been demonstrated by direct observations [76], the existence of a sheet for streamers could until recently only be inferred from the shape of these streamers and their association with photospheric magnetic fields. Indirect evidence could come from radar observations of the sun [91, 92], which can in principle locate a current sheet as a region of pronounced plasma turbulence. It has been shown [73] that if the frequency of Coulomb collisions is small a thin current sheet is unstable against the excitation of ion-acoustic waves, and the appearance of such a sheet must be associated with the onset of plasma turbulence.

Recent observational data can be taken as direct evidence for the existence of current sheets in streamers. Study of the slowly varying component of the solar radio emission and comparison of this component with the visible optical features on the solar disc [93] have shown that these streamers make an important contribution to the thermal radio emission of the sun. Furthermore, observations have been used to work out a morphological model for streamers, and this model agrees well with the model which we have been discussing in this chapter.

§ 3. Some Comments Regarding Two-Dimensional Models of the Magnetosphere

Mathematical models of actual phenomena always involve certain idealizations. Fortunately, the assumptions which we are using in this analysis of two-dimensional models of the magnetosphere are not so crude as to fundamentally alter the physical mechanism of the phenomenon. Furthermore, near a current sheet, which can be treated as a surface, the two-dimensional approach is natural and common. Finally, experience shows that study of two-dimensional models is extremely useful and can reveal several important features of the geomagnetic field [94, 95] (Figs. 13a and 13b).

Furthermore, two-dimensional models can be used for concrete quantitative calculations, e.g., in connection with laboratory experiments on flow around a plane magnetic dipole [24-26].

A systematic approach to the problem of the stretching out of the geomagnetic tail by plasma flow [96-100], even for a two-dimensional geometry, requires solution of a mathematically complicated transient problem. The magnetic field of the plane dipole initially fills some cylindrical cavity in the plasma (Fig. 13c). Then the solar wind is turned on. The incoming plasma flow deforms the boundary of the cavity and captures part of the magnetic flux. If this capture did not occur the cavity would be unbounded in one direction (Figs. 13a and 13b) only if the gas pressure of the plasma p_0 vanished. Since this is not the case (in fact, this pressure is quite high), the geomagnetic cavity must be closed ([96]; see also [97]).

The front (day) side of the boundary acts as a compressing wall, while the back (night) side acts as a wall extracting the field, in the case in which the field is captured by the wind. This boundary serves as the separation boundary in the problem treated in the preceding section (Fig. 13d). As time elapses a current sheet again forms, and the steady-state shape of the geomagnetic cavity is established. Near the current sheet the magnetic field has a structure similar to that of the asymptotic solution (Fig. 12), while the lateral boundaries are governed by the equilibrium between the magnetic field in the cavity and the plasma of the solar

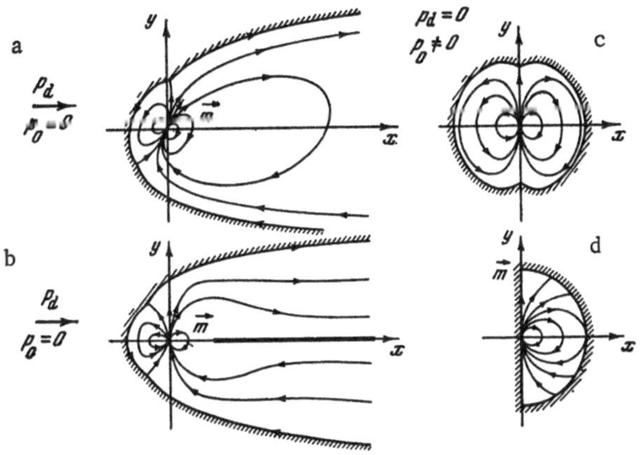

Fig. 13. Two-dimensional model of the magnetosphere. a) According to Zhigulev and Romishevskii [94]. The shape of the boundary is governed by the equality of the dynamic pressure P_d in the plasma flow and the magnetic pressure within the magnetosphere; b) according to Unti and Atkinson [95]. The effect of the current sheet corresponding to the geomagnetic tail on the shape of the magnetospheric boundary is taken into account; c) two-dimensional cavity in a plasma [96] filled by the magnetic field of a plane dipole. At the boundary of the cavity the magnetic pressure is equal to the static pressure P_0 of the surrounding plasma; d) configuration of the magnetic field of a plane dipole with lines of force frozen at the boundary of the semicylindrical region.

wind. The equilibrium cavity shape was found by Unti and Atkinson [95] for a symmetric two-dimensional formation of the problem with a current sheet. In general, with a two-dimensional dipole inclined at some angle with respect to the plasma flow, the shape of the current sheet itself is not known. Oberts determined the steady-state shape of the current sheet [101]. Our model shows how and why a configuration containing a current sheet arises:

(1) The geomagnetic tail forms as a result of the capture of the geomagnetic field by plasma of the solar wind.

(2) The currents induced at the surface of the magnetosphere distort the original field so that singular null points (or lines) would appear in the absence of plasma within the magnetosphere.

(3) A current sheet develops at the singular null points in the plasma and produces the field structure corresponding to the geomagnetic tail.

Let us examine the conditions listed earlier for the appearance of a current sheet in connection with the problem of the magnetospheric tail [98].

The first of these conditions – the existence of a region with a strong field satisfying (2.1) for the region of the magnetic tail – is obviously satisfied, since the field in the magnetic tail out to the lunar orbit is at least 10γ, while the plasma density is $n \ll 1\ cm^{-3}$ and the temperature is $T \lessdot 10^6$ °K. The plasma velocity is $V \ll V_A \sim 10^8$ cm sec (here we are not talking about the rapid transient processes which occur in the magnetosphere).

The second condition – the existence of a boundary separating the strong-field region in the geomagnetic cavity from the wind region, in which the dynamic plasma pressure dominates – is also obviously satisfied. In the case of the magnetosphere this boundary is the magnetopause.

The third condition – regarding the capture of the geomagnetic field by the solar wind – is of fundamental importance and was discussed in [98, 100]. Obviously, if we assume that the plasma has an ideal conductivity, a current sheet, once it appears, will exist indefinitely and need not be maintained by any external forces. In particular, this assertion means that the magnetic lines of force are everywhere parallel to the boundary surface, except the infinitely remote trailing part of this boundary at which the field is captured.

Actually, however, there is always some field dissipation. It also occurs in a collisionless plasma, as the result of instabilities developing at a current sheet. Then if a steady or quasisteady shape of the geomagnetic tail is to be maintained, we require an electromotive force which acts at all times and maintains the current in the sheet.

This electromotive force is probably of an induction nature, due to some dissipation of magnetic flux at the lateral boundary of the geomagnetic tail (and, in general, at the magnetopause) [98, 100]. At the same time such dissipation gives us a natural explanation for the capture of the geomagnetic field by the solar wind. We assume that at the lateral surface of the tail there is some small normal field component $H_\perp \ll H_\parallel$, where $H_\parallel = H$ is the field in the tail (Fig. 14). The presence of $H_\perp$ automatically leads to the appearance of an induced electromotive force. Let us consider the generalized Ohm's law:

$$\mathbf{j} = \sigma\left(\mathbf{E} + \frac{1}{c}[\mathbf{vH}]\right) \tag{2.36}$$

(here we are neglecting the gradient terms in a first approximation, even though these terms can turn out to be important, e.g., in interpreting the plasma sheet in the magnetospheric tail). Here σ is the conductivity, which we assume to be infinite everywhere except at a current sheet of length L, width $2y_c$, and thickness $2z_c$; $\mathbf{v}$ is the velocity of the solar wind. Integrating Eq. (2.36) over a contour around half the tail cross section (Fig. 14), we find, for the steady state (rot E = 0),

$$\oint \frac{\mathbf{j}}{\sigma} d\mathbf{l} = \frac{1}{c} \oint [\mathbf{vH}]\, d\mathbf{l},$$

or

$$j = \pi\sigma \frac{V}{c} H_\perp, \tag{2.37}$$

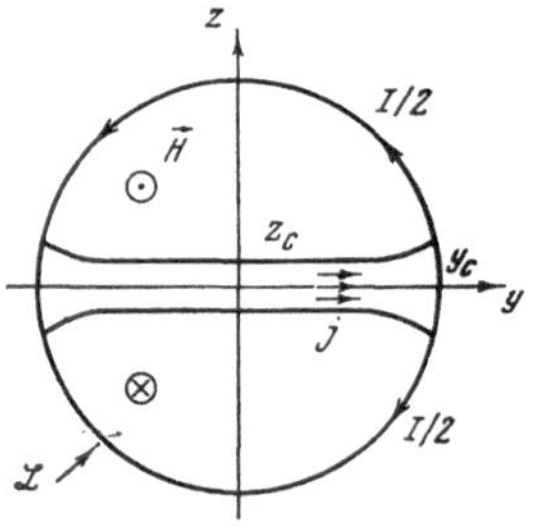

Fig. 14. Schematic cross section of the geomagnetic tail in a plane perpendicular to the earth-sun axis (the x axis).

where we have used the approximation that the perimeter of the tail cross section is $2\pi y_c$. On the other hand, the current density at the sheet can be estimated from

$$j \simeq \frac{I}{2z_0} = \frac{cH_\parallel}{4\pi z_0}, \tag{2.38}$$

and conservation of magnetic flux requires

$$LH \approx y_0 H_\parallel. \tag{2.39}$$

Then from Eqs. (2.37) and (2.38) we find

$$L = \frac{4\pi^2\sigma}{c^2} V y_0 z_0. \tag{2.40}$$

If we adopt as the conductivity σ the Buneman value $\sigma = (1/2)(M/m)^{1/3}\omega_{0e} \simeq 3.4 \cdot 10^5 \sqrt{n}$ [102], then with a density of $n \simeq 10$ in the sheet we would have $\sigma \simeq 10^6$. Also assuming $V = 3 \cdot 10^7$ cm/sec, $y_c \simeq 20R_E \simeq 1.2 \cdot 10^{10}$ cm, and $z_c \simeq 0.01$-$0.1R_E$, we find

$$L \simeq 10^4 z_0 \simeq (10^2 - 10^3)\, R_E, \tag{2.41}$$

where $R_E = 6 \cdot 10^8$ cm is the earth's radius. Since the Buneman value of the conductivity is probably a lower limit, and the actual conductivity can be far higher, the estimate of the tail length in (2.41) should also be treated as a lower limit.

§ 4. Magnetic Field of a Contracting Plasma Cylinder

A. Magnetic Field Configuration in the Absence of a Plasma

One of the old problems of cosmic electrodynamics is that of the contraction of a gravitating plasma cloud with a magnetic field which couples the cloud to the surrounding medium [103]. In discussions of various aspects of this problem particular attention has been paid to the configuration of the magnetic field of the cloud and its time evolution. The most detailed results in this direction for a cloud with an extended inhomogeneous atmosphere were obtained in [104, 105].

Let us consider the simplest case: that in which there is no plasma at all outside the cloud. Then if the cloud is formed in a uniformly magnetized medium with an initial field H_0, which extends to infinity, the solution of the problem for the magnetic field outside the cloud is obvious [103, 15]. The magnetic field is axisymmetric (the z axis is parallel to H_0) and is the superposition of a dipole field and the initial uniform field:

$$\Phi(r, \theta, t) = \frac{1}{2} H_0 r^2 \left[1 + \frac{R}{R_0}\left(1 - \frac{R^2}{R_0^2}\right)\left(\frac{R_0}{r}\right)^3\right] \sin^2\theta. \tag{2.42}$$

Here $\Phi(r, \theta, t)$ is the stream function, defined by (1.47), $R = R(t)$ is the running value of the cloud radius, and R_0 is the initial cloud radius. Figure 15 shows the pattern of lines of magnetic force.

A characteristic structural feature of the magnetic field is a null line of the X type, a circle in the $\theta = \pi/2$ plane of radius

$$R_x = R_0\left[\frac{R}{2R_0}\left(1 - \frac{R^2}{R_0^2}\right)\right]^{1/3}. \tag{2.43}$$

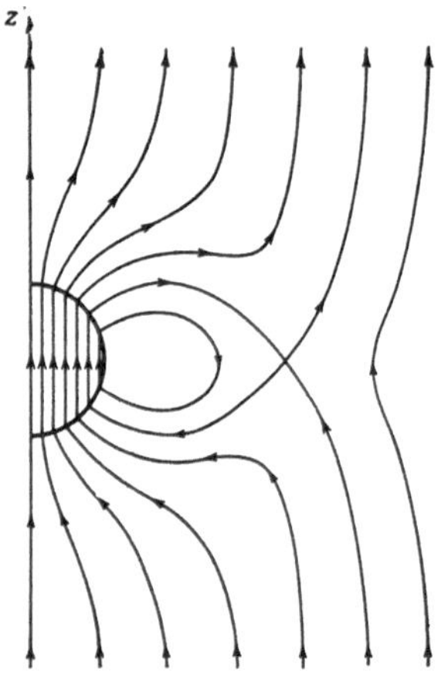

Fig. 15. Magnetic lines of force in the meridional plane outside a contracting cloud.

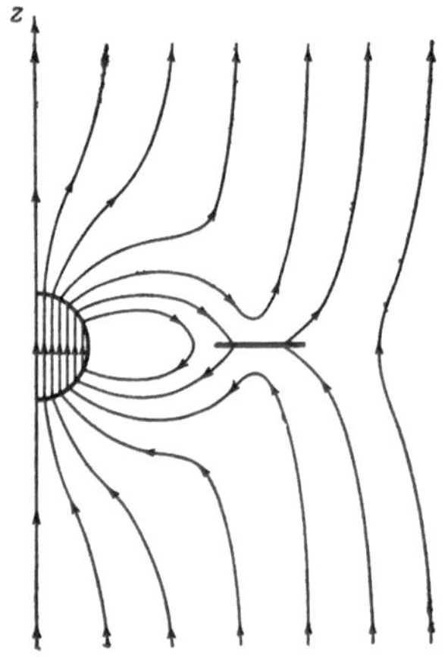

Fig. 16. Magnetic lines of force outside a contracting cloud for the case in which there is a current sheet (ring) which has developed from a null line.

The null line appears at the equator of the cloud, at $R = R_0/\sqrt{3}$. At this time the plasma (even at a low density) outside the cloud significantly alters the entire field pattern: A current begins to flow near the null line. It stretches out into a thin current sheet or a current ring around the cloud. In the general case of a finite conductivity the dissipative processes in the current sheet can lead to a separation of the current ring from the cloud, and the magnetic field becomes qualitatively like that shown in Fig. 16. However, it is a difficult matter to find an analytic solution for the corresponding three-dimensional problem, so we restrict the analysis to the following formulation: We seek the magnetic field of a contracting plasma cylinder surrounded by an ideally conducting plasma of low density with a magnetic field which becomes uniform at large distances.

In the absence of an external plasma the solution in terms of the complex potential in (1.14) is

$$F(z, t) = -H_0 z - (H_R - H_0)\, R^2/z. \tag{2.44}$$

Here H_0 is the initial uniform field, parallel to the y axis, and $H_R = H_0 R_0/R$ is the uniform field frozen in the plasma in the contracting cylinder. The pattern of lines of force in the (x, y) plane is the same as for the three-dimensional (axisymmetric) case in the meridional plane (Fig. 15). However, the null line (two lines parallel to the z axis) now appears at the magnetic equator of the cylinder at $R = R_0/2$. Accordingly, if we wish to take into account the role of the plasma outside the cylinder in the approximation of a strong field and an infinite conducti-

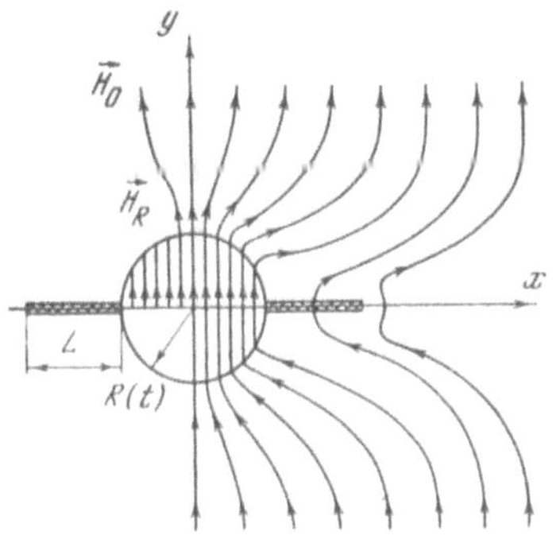

Fig. 17. Magnetic field of a contracting cylinder (the general case).

vity, then with $R < R_0/2$ we must introduce two cuts in the complex plane, corresponding to the current sheets developing from the null line, as shown in Fig. 17.

B. Formulation and Solution of the Problem with a Current Sheet

The magnetic field is a potential field with

$$\Delta A = 0 \tag{2.45}$$

outside a circle of radius R with two cuts ($y = 0$, $R \le |x| \le R + L$) of length L. It satisfies the boundary conditions

$$A = \begin{cases} -H_R R \cos\varphi & \text{for} \quad |z| = R, \\ -H_R R & \text{for} \quad y = 0, \quad R \leqslant x \leqslant R + L, \\ H_R R & \text{for} \quad y = 0, \quad -R - L \leqslant x \leqslant -R, \end{cases} \tag{2.46}$$

$$A \to -H_0 x \quad \text{for} \quad |z| \to \infty. \tag{2.47}$$

The problem is solved by a method completely analogous to that of §2 in this chapter. The solution is

$$F(w, t) = F_0(w, t) + F_1(w, t) + F_2(w, t), \tag{2.48}$$

where

$$F_0(w, t) = -H_0 z(w) = -\frac{H_0 R}{2\cos\beta_0}\left\{\left(w + \frac{1}{w}\right) + \left[\left(w + \frac{1}{w}\right)^2 - 4\cos^2\beta_0\right]^{1/2}\right\}, \tag{2.49}$$

$$F_1(w, t) = \frac{2H_0 R}{\cos\beta_0} w - \frac{H_0 R}{2\cos\beta_0}\left\{\left(w + \frac{1}{w}\right) - \left[\left(w + \frac{1}{w}\right)^2 - 4\cos^2\beta_0\right]^{1/2}\right\}, \tag{2.50}$$

$$F_2(w, t) = -\frac{H_R R w}{\cos\beta_0} - \frac{H_R R}{\cos\beta_0}\frac{2}{\pi}\left[\cos\beta_0 \frac{i}{2}\ln\frac{1 - w^2 - 2iw\sin\beta_0}{1 - w^2 + 2iw\sin\beta_0} + \right.$$
$$\left. + \frac{i}{4}\left(w + \frac{1}{w}\right)\ln\frac{1 - w^2\exp(-2i\beta_0)}{1 - w^2\exp 2i\beta_0} - \beta_0 w\right], \tag{2.51}$$

$$w(z, t) = \frac{\cos\beta_0}{2}\left\{\left(\frac{z}{R} + \frac{R}{z}\right) - \left[\left(\frac{z}{R} + \frac{R}{z}\right)^2 - \frac{4}{\cos^2\beta_0}\right]^{1/2}\right\}, \tag{2.52}$$

$$\cos\beta_0 = \frac{2(1 + \xi)}{2(1 + \xi) + \xi^2}, \qquad \xi = \frac{L}{R}. \tag{2.53}$$

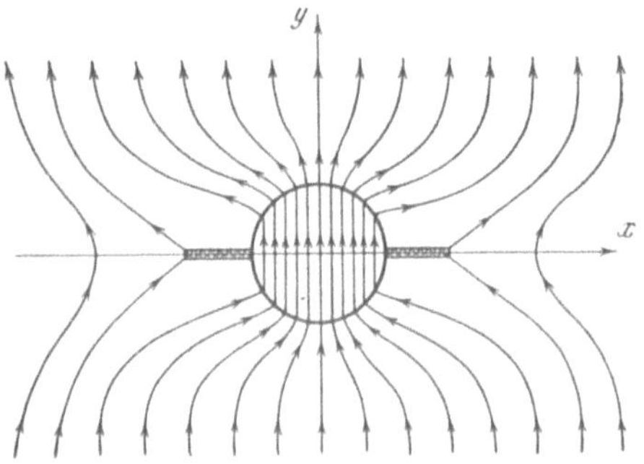

Fig. 18. Magnetic lines of force in the particular case in which there is no return currents (in which the cut has a minimum length).

As in the preceding problem (§2 of Chapter II), the general solution falls between two limiting cases: (1) that in which there are no return currents in the sheet and (2) that in which the total current in the sheet vanishes.

The first case corresponds to the minimum cut length:

$$L = L_1 = R\left(\cot\frac{\pi R}{R_0} - 1 + \operatorname{cosec}\frac{\pi R}{R_0}\right). \tag{2.54}$$

In the second case we have

$$L = L_2 = R\,(\tan\beta_{02} - 1 + \sec\beta_{02}), \tag{2.55}$$

where β_{02} is the root of the transcendental equation

$$\beta_{02} = \frac{\pi}{2}\left(1 - \frac{2R}{R_0}\right) - \frac{1}{2}\cot\beta_{02}\cdot\ln\cos\beta_{02}. \tag{2.56}$$

The pattern of magnetic lines of force in the general case is the same as in the case in which the total current vanishes; it is shown schematically in Fig. 17. In the particular case in which there are no return currents the lines of force have the pattern shown in Fig. 18.

Accordingly, when a contracting cylinder or sphere is formed from a uniformly magnetized plasma the motion of the plasma leads to the formation of a current sheet with the corresponding magnetic field configuration. In the case of a sphere the current sheet is a thin equatorial current ring, while in the case of a cylinder it consists of two current sheets, which require a closing contour at infinity.

§ 5. Three-Dimensional Problem with a Current Sheet

The example of the preceding section demonstrates the obvious relationship between the three-dimensional axisymmetric problem and the two-dimensional problem of a current sheet which develops from a null line as a cylinder contracts in a magnetized plasma. Using the rule proposed in [19] for constructing current sheets in the presence of magnetic null lines, we turn now to more complicated questions.

What shape would we expect for a current sheet in, for example, real coronal streamers? To answer this question even qualitatively we must know what shape the magnetic null line would have if there were no plasma between two conducting boundaries, at one of which there is a three-dimensional dipole. We assume for simplicity that these boundaries are planar and ideally conducting, and we treat the following simple three-dimensional problem.

We assume a three-dimensional dipole, oriented along the x axis and lying at the origin in the (x, z) plane (Fig. 19). We have the photosphere in mind as the ideally conducting (x, z)

Fig. 19. Diagram used to formulate the three-dimensional problem of the appearance of a current sheet as the magnetic field of an axisymmetric dipole **m** is stretched out by a flat, ideally conducting wall S.

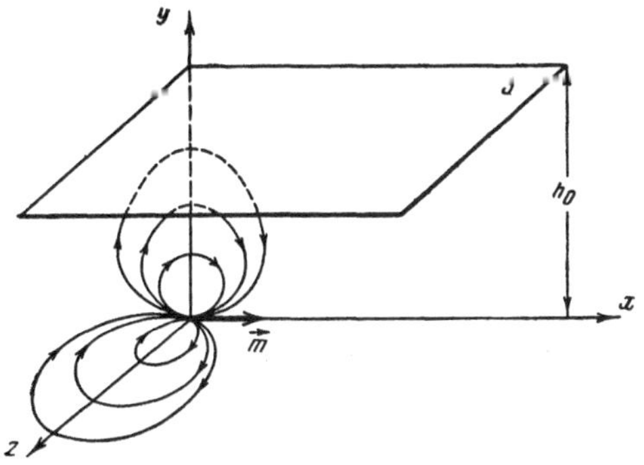

boundary plane. We assume that the magnetic field of the dipole is frozen in the other ideally conducting plane boundary S, which is initially at a height h_0 above the (x, z) plane. We assume that this boundary is subsequently raised to a height $h > h_0$. What is the magnetic field within the plane slab between these ideally conducting boundaries? This field is obviously governed by the internal Neumann problem:

$$\Delta B = 0, \tag{2.57}$$

$$\left.\frac{\partial B}{\partial y}\right|_{y=0} = 0, \tag{2.58}$$

$$\left.\frac{\partial B}{\partial y}\right|_{y=h} = \frac{3mh_0 x}{(x^2 + h_0^2 + z^2)^{5/2}}. \tag{2.59}$$

Here B is the magnetic scalar potential:

$$\mathbf{H} = \nabla B. \tag{2.60}$$

For a point dipole we would have

$$B = -\mathbf{mr}/r^3. \tag{2.61}$$

It is not difficult to find a solution for this problem. It can be written as the rapidly converging series

$$B = -mx/r^3 + mx \sum_{n=0}^{\infty} \{[x^2 + z^2 + (y - h_1 - 2nh)^2]^{-3/2} + [x^2 + z^2 + (y + h_1 + 2nh)^2]^{-3/2}\}, \tag{2.62}$$

where

$$h_1 = h + hh_0 (h^4 - h_0^4)^{-1/4}. \tag{2.63}$$

The first term in the series (n = 0),

$$B^{(0)} = -mx/r^3 + mx \{[x^2 + z^2 + (y - h_1)^2]^{-3/2} + [x^2 + z^2 + (y + h_1)^2]^{-3/2}\}, \tag{2.64}$$

describes a potential magnetic field having an X-type null line, i.e., a closed curve in the (y, z) plane described by the equation

$$(y^2 + z^2)^{-3/2} - [(y - h_1)^2 + z^2]^{-3/2} - [(y + h_1)^2 + z^2]^{-3/2} = 0. \tag{2.65}$$

Because of the way the problem is formulated, it is sufficient to consider only the upper half of this curve (Fig. 20). The principal semiaxes of the curve for $R \geq 2^{1/2} h_0$ are

$$a = h_1 (2^{2/5} - 1)^{-1/2} \simeq 1.3 h_1, \qquad b \simeq 0.494 h_1. \tag{2.66}$$

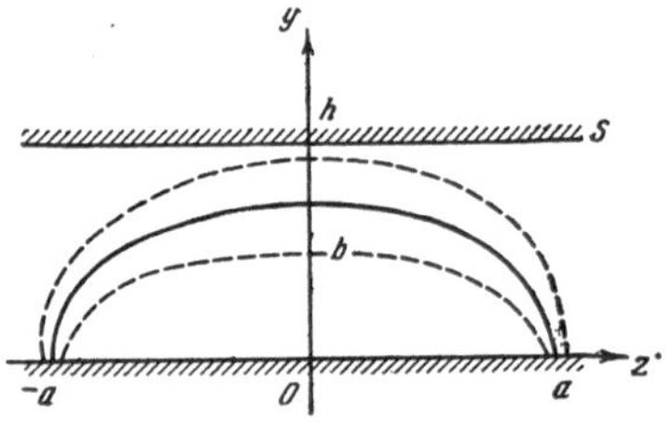

Fig. 20. Null line which appears when a magnetic dipole field is stretched out by a flat conducting wall S; the current sheet which develops from it (dashed curves).

We now assume that there is a plasma in this plane slab between the conducting walls and that the conditions corresponding to the strong-field approximation, (14), hold. Then as h increases, a current flows along the null line. Using the rule proposed in [19] we can easily see that, because of the interaction with the magnetic field, this current "spreads out" into a thin current sheet, in a crescent whose ends are resting on the photosphere. The boundaries of the current sheet are shown schematically by the dashed curves in Fig. 20.

These simple considerations demonstrate one possible closing of the current sheet in the lower part of a coronal streamer – closing through the photosphere. A decisive test of this hypothesis might come from a study of that component of the magnetic field at the base of a streamer which is tangent to the photosphere.

In the upper part of a streamer we believe it would be more probable to find a closing of the current sheet analogous to that which occurs in the geomagnetic tail (Fig. 14).

In its general three-dimensional formulation the problem of the shape of a current sheet is complicated and has not yet been solved. A solution would require a generalization of the principles set forth in [19]. The concrete shape of this generalization will of course depend on the symmetry of the problem. As was shown in §4 of this chapter, there is no difficulty in transforming from symmetric two-dimensional problems with current sheets to three-dimensional axisymmetric problems. However, even asymmetric two-dimensional problems, e.g., the problem of plasma flow around an inclined plane dipole [101], evidently require additional equations for their solution – equations describing a minimum of the magnetic energy, the pressure balance at the current sheet, etc.

A constituent part of the overall problem is that of the topology of three-dimensional fields with null points, lines, etc. A local classification of three-dimensional null points was given in [106] in an effort to solve this part of the problem. The topology of static magnetic fields is treated in [107]. In an extremely useful study [108] (see also [109]), Sweet used the example of a magnetic field symmetric about a plane (an ideally conducting photosphere) to show that the role of the null line can be played by a line analogous to a null line in the sense that it separates magnetic fluxes, even though there is no component of the magnetic field along it. It is along this line that a current should flow and nucleate the current sheet upon changes in a magnetic field accompanied by a redistribution of magnetic flux.

CHAPTER III

HYDRODYNAMIC PLASMA FLOW NEAR A CURRENT SHEET

Important progress has now been made toward an understanding of the fundamental role of null points (or lines) of a magnetic field in plasma dynamics, especially in applications in space physics (see the discussion above as well as [19, 75] and the literature cited there). The general thrust of the research has now turned toward the analysis of the concrete physical processes which occur in a plasma near such points [19, 110, 114].

Syrovatskii [19] studied the question of the structure of a strong magnetic field frozen in a plasma in the case of two-dimensional flow. He showed that the presence of a neutral (null) line of a magnetic field along which an electric field is directed is a necessary and sufficient condition for the appearance of a current sheet. In general the current sheet which develops consists of forward and return currents; there are upper and lower limits on the magnitude of the return current. In one limit, the total current in the sheet may be matched with the external field in such a manner that there is no return current at all. In this case the field structure corresponds to that found previously for self-similar solutions [110, 111]. In the other limit the total current in the sheet vanishes; i.e., the return current completely balances the forward current. Such a current sheet need not be electrically closed. In [19] Syrovatskii solved that part of the overall problem which deals with the magnetic field configuration near a current sheet which arises at an existing null line.

Below we continue the study of plasma flow near a simple null line, making use of the results of [19]. Specifically, we now turn to the second part of the problem – that of calculating the profiles of the plasma velocity and density and their time dependence. One of the main results is a rigorous proof of the assertion in [16, 17] that a pronounced decrease in plasma pressure occurs as a plasma moves near a null line.

§ 1. General Formulation of the Problem

As before, we use the approximation of a strong field and a cold plasma, starting from Eqs. (1.19)-(1.22). In this approximation the magnetic field is a potential field, except perhaps at certain surfaces [cuts in the (x, y) plane], along which currents flow (i.e., except for current sheets). Except at these singularities the potential magnetic field, evolving over time in accordance with the boundary conditions (which deal with the external sources and current sheets), leads to plasma motion. The kinematics of this motion is unambiguously governed by two conditions: the condition that the acceleration be orthogonal to the magnetic lines of force, (1.20), and the frozen-in condition, (1.21). We assume that x(t) and y(t) are the coordinates of some "fluid particle." Then according to (1.20) and (1.21) we can write

$$\ddot{x}A'_y - \ddot{y}A'_x = 0, \tag{3.1}$$

$$\dot{x}A'_x + \dot{y}A'_y + A'_t = 0. \tag{3.2}$$

Here the dot denotes differentiation with respect to the time t, while the prime denotes differentiation with respect to the variable given in the subscript. Differentiating (3.2) with respect to the time, we find from (3.1) and (3.2) a system of four first-order differential equations:

$$\begin{aligned} &\dot{x} = u, \qquad \dot{y} = v, \\ &\dot{u} = -\frac{A'_x}{A'^2_x + A'^2_y} f(t, x, y, u, v), \\ &\dot{v} = -\frac{A'_y}{A'^2_x + A'^2_y} f(t, x, y, u, v), \end{aligned} \tag{3.3}$$

where

$$f(t, x, y, u, v) = u(A''_{xx}u + A''_{xy}v + A''_{xt}) + v(A''_{xy}u - A''_{xx}v + A''_{yt}) + (A''_{xt}u + A''_{yt}v + A''_{tt}). \tag{3.4}$$

Supplemented by the initial values of the coordinates and velocity of the fluid particle,

$$x(0) = x_0, \quad y(0) = y_0, \quad u(0) = u_0, \quad v(0) = v_0, \tag{3.5}$$

the system of equations (3.3) unambiguously determines the trajectory of this particle at subsequent times. We recall that u_0 and v_0 are related by condition (1.26). Accordingly, if we assume that the plasma is initially (at t = 0) at rest,

$$x(0)=x_0,\quad y(0)=y_0,\quad u(0)=0,\quad v(0)=0, \tag{3.6}$$

then we must require that

$$A_t'=0 \quad \text{at} \quad t=0. \tag{3.7}$$

The change in the density of the fluid particle as it moves along this trajectory is governed by the continuity equation (1.22), rewritten in Lagrange form (1.38).

We turn now to an analysis of the plasma motion in the field of a current sheet which arises at the lowest-order null line (n = 2; [19]), which is at the origin of the (x, y) plane. The complex potential of such a sheet is [19]

$$F(z,\ t)=\frac{\alpha}{2}z\sqrt{z^2-b^2}+\frac{\Gamma}{2\pi}Ln\frac{z+\sqrt{z^2-b^2}}{b}+A(t). \tag{3.8}$$

Here α is the magnetic field gradient near the initial null line, b = b(t) is the half-width of the current sheet at time t, $\Gamma = -\pi\alpha b^2\nu$ is proportional to the total current in the sheet, and $\nu = \nu(t)$ is a dimensionless parameter, whose possible values are generally governed by t and lie between zero (in the case in which the total current vanishes) and unity (in the case in which there is no return current). Also, A(t) is the magnetic flux which is annihilated at the current sheet at time t.

Figure 21 shows the pattern of lines of force A(x, y, t) = const before the current sheet appears (Fig. 21a) and after it has appeared, in the two limits defined above (Figs. 21b and 21c).

Typically, the magnetic field near a current sheet has null points (or lines) at the surface of the sheet (Figs. 21b and 21c). The coordinates of these points, determined from the condition $\partial F/\partial z = 0$, are

$$x_*=\pm\left(\frac{b^2}{2}-\frac{\Gamma}{2\pi\alpha}\right)^{1/2}=\pm b\left(\frac{1+\nu}{2}\right)^{1/2},\qquad y_*=\pm 0. \tag{3.9}$$

As was mentioned above, these null points differ from the original null points (Fig. 21a) in that they correspond to nonanalytic singularities at the boundary of the region of applicability of the solution – at the cut in the complex plane. The expansion of the potential near such singularities is

$$F(z,\ t)\simeq\pm\frac{C_1}{2}(z-z_*)^2+iC_2,\qquad y_*=\pm 0, \tag{3.10}$$

where $z_* = x_* + iy_*$; C_1 and C_2 are time-dependent constants; and y = +(–)0 corresponds to the upper (lower) edge of the cut.

An exceptional case is the limit $\nu = 1$, in which case the null points coincide with the ends of the cuts, $x_* = \pm b$, and the potential expansion is

$$F(z,\ t)\simeq d_1(z-x_*)^{3/2}+d_2, \tag{3.11}$$

where d_1 and d_2 are time-dependent constants.

The minimum value of $|x_*|$ for a given value of b is found in the second limiting case, $\nu = 0$, for which we have

$$x_*=b/\sqrt{2}. \tag{3.12}$$

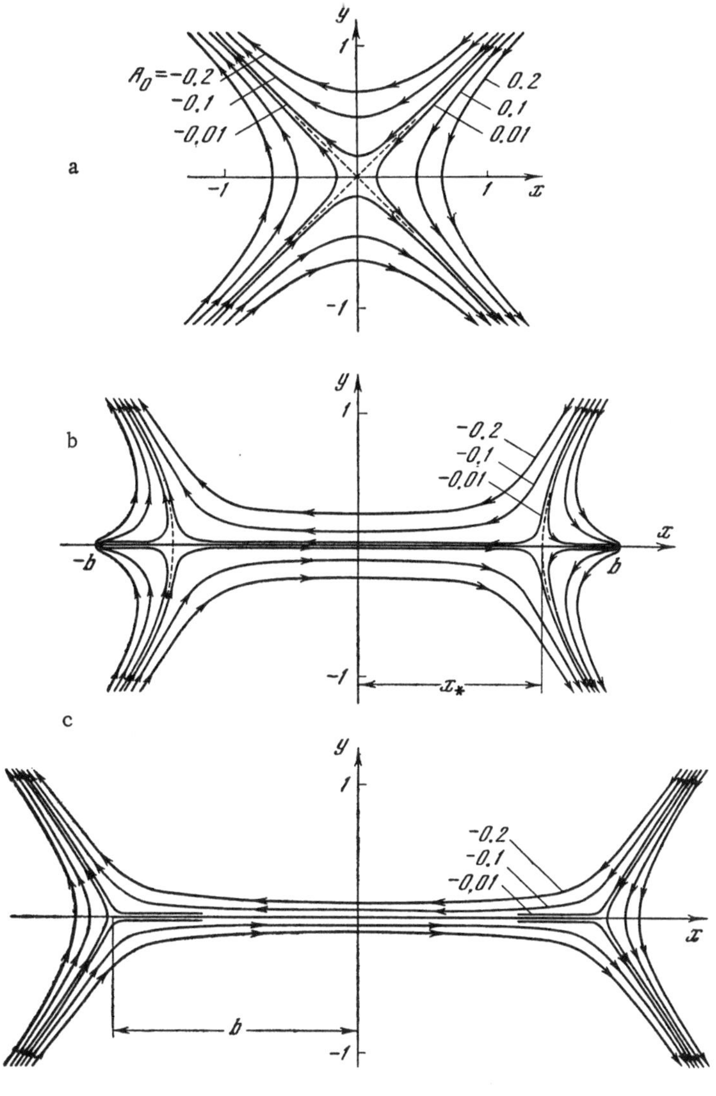

Fig. 21. Lines of force of the magnetic field of a developing current sheet in the case A(t) = 0. a) Lines of force of a first-order null point at t = 0. The curves are labeled with the values of the vector potential $A(x_0, y_0, 0) \equiv A_0$; b) the same lines of force at t = 1 for the case of a current sheet which has developed from a null point. The limiting case in which the total current in the sheet vanishes, $\nu = 0$, is shown. Here $x_* = b/2^{1/2}$ gives the position of the null points on the surface of the sheet ($b \simeq 2$, t = 1); c) the same, in the case in which there is no return current, $\nu = 1$. The null points are at the ends of the current sheet ($b \simeq 2$, $t \simeq 5.6$).

As was shown in [19] the half-width b(t) of the current sheet is governed by the parameters of the external field, the total current in the sheet, and the quantity A(t) through the equation

$$\frac{\alpha}{4} b^2 + \frac{\Gamma}{2\pi} \ln \frac{b}{2} - \frac{\Gamma}{2\pi} L_0 - A(t) - \beta(t) = 0. \qquad (3.13)$$

Here β (t) is a parameter which is a measure of the external electric field; this field is parallel to the null line, $E = -c^{-1}\partial A/\partial t$, and is responsible for the appearance of the current sheet from the null line and its growth. Here A(t) is a measure of the rate of dissipation of the magnetic field in the current sheet.

Below we consider two problems: (1) that of hydrodynamic plasma flow near a developing current sheet; (2) flow near a steady-state current sheet of constant width 2b. In the first problem, which corresponds to the stage in which the current sheet is being formed, we will neglect the magnetic field dissipation at the current sheet altogether; i.e., we will assume

$$A(t) \equiv 0. \qquad (3.14)$$

In the second problem, in contrast, we assume that this rate of dissipation is such that the sheet width is maintained at a constant value; specifically we assume

$$A(t) = A(0) - \beta(t), \qquad (3.15)$$

where

$$A(0) = \frac{\alpha}{4} b^2 + \frac{\Gamma}{2\pi} \ln \frac{b}{2} - \frac{\Gamma}{2\pi} L_0 \qquad (3.16)$$

and b = const.

§ 2. Hydrodynamic Plasma Flow near a Developing Current Sheet

We assume that the plasma is initially at rest and that the strong frozen-in magnetic field is hyperbolic (Fig. 21a). At this time a constant and uniform electric field $\mathbf{E}_0$ is applied (with a small switching time τ). This field is parallel to the magnetic null line. As β(t) we adopt the function

$$\beta(t) = cE_0 t^{k+1}/(\tau + t)^k, \qquad (3.17)$$

where k is a positive integer. Here we have $\beta(0) = 0$, which ensures that assumption (3.7) is valid.

As was mentioned above, we set the function A(t) in Eq. (3.13) identically equal to zero. The constant $L_0 = \oint dl/r$ is governed by the geometry of the current loop closing the current sheet: The integration is carried out along this contour so that we have $L_0 \simeq \ln L_n$, where L_n, the scale dimension of the closing contour, is obviously much larger than b, except in the case $\nu = 0$, in which this contour is not necessary at all.

This problem of the particle trajectory in a potential force field (the size of the field itself is not important, since only the acceleration direction appears in the equation of motion) is purely kinematic, as can be seen in the circumstance that the problem has only two dimensional parameters—the scale length L and the scale time T. Choosing $L = (cE_0T/\alpha)^{1/2}$, we find a dimensionless form of Eq. (3.13):

$$\frac{b^2}{4} - \nu \frac{b^2}{2} \ln \frac{b}{2l} = \frac{t^{k+1}}{(\tau + t)^k}, \qquad (3.18)$$

where the parameter $l = L_n/L$ is much larger than unity by virtue of the arguments above. Solutions for the case $l = \bar{l}$ were given in [114]; these solutions hold only for small values of b.

To calculate the right sides of the system of differential equations in (3.3) using current-sheet potential (3.8) it is convenient to differentiate the complex potential F(z, t). For example, the derivative

$$\frac{\partial F(z,t)}{\partial z} = A'_x - iA'_y = \alpha\left[(z^2 - b^2)^{1/2} - \lambda b^2 (z^2 - b^2)^{-1/2}\right], \tag{3.19}$$

where $\lambda = (1 - \nu)/2$, yields

$$A'_x = \alpha\left(q + \frac{\lambda b^2}{q}\right)\cos\psi, \tag{3.20}$$

$$A'_y = -\alpha\left(q - \frac{\lambda b^2}{q}\right)\sin\psi. \tag{3.21}$$

Here q and ψ are the modulus and argument of $(z^2 - b^2)^{1/2}$.

Differentiating the complex potential F(z, t) with respect to the time and using (3.14) we find

$$\frac{\partial F}{\partial t} = \frac{\partial F}{\partial b}\dot{b} = -\alpha\left[\lambda\frac{z}{\sqrt{z^2 - b^2}} + \nu\ln\frac{z + \sqrt{z^2 - b^2}}{b}\right]b\dot{b}. \tag{3.22}$$

Hence

$$A'_t = -\alpha\left[\lambda\frac{r}{q}\cos(\varphi - \psi) + \nu\ln Q - \nu\ln b\right]b\dot{b}, \tag{3.23}$$

where

$$Q = \left|z + \sqrt{z^2 - b^2}\right|. \tag{3.24}$$

Similarly, we can calculate the second derivatives:

$$A''_{xx} = \alpha\left[\frac{r}{q}\cos(\varphi - \psi) - \lambda b^2\frac{r}{q^3}\cos(\varphi - 3\psi)\right], \tag{3.25}$$

$$A''_{xy} = -\alpha\left[\frac{r}{q}\sin(\varphi - \psi) - \lambda b^2\frac{r}{q^3}\sin(\varphi - 3\psi)\right], \tag{3.26}$$

$$A''_{xt} = \alpha\left[\frac{\lambda b^2}{q^3}\cos 3\psi - \frac{\nu}{q}\cos\psi\right]b\dot{b}, \tag{3.27}$$

$$A''_{yt} = \alpha\left[\frac{\lambda b^2}{q^3}\sin 3\psi - \frac{\nu}{q}\sin\psi\right]b\dot{b}, \tag{3.28}$$

$$A''_{tt} = -\alpha\left[\lambda\frac{r}{q}\cos(\varphi - \psi) + \nu\ln Q - \nu\ln b\right](b\ddot{b} + \dot{b}^2) - \\ - \alpha\left[\lambda\frac{r}{q^3}\cos(\varphi - 3\psi) - \frac{\nu}{b^2}\frac{r}{q}\cos(\varphi - 3\psi)\right](b\dot{b})^2. \tag{3.29}$$

Here

$$z = re^{i\varphi}. \tag{3.30}$$

The quantity b at any time is defined as the root of Eq. (3.18); its derivatives are

$$\dot{b} = \beta\left[\alpha b(\lambda - \nu\ln b + \nu\ln 2l)\right]^{-1}, \tag{3.31}$$

$$\ddot{b} = \frac{\dot{\beta}/\alpha - \dot{b}^2(\lambda - \nu - \nu\ln b + \nu\ln 2l)}{b(\lambda - \nu\ln b + \nu\ln 2l)}. \tag{3.32}$$

Accordingly, the problem of plasma motion in the magnetic field of a developing current sheet reduces to the integration of the system of nonlinear differential equations in (3.3) with initial conditions (3.6). The right sides of Eqs. (3.3) depend on the derivatives of the vector potential A(x, y, t). This vector potential is the real part of the complex potential in (3.8), whose time dependence is governed by the parameter b(t), which is found from Eq. (3.18). In this equation we assume ν = const. Furthermore, to find the plasma-density profile we must calculate the Jacobian in (1.38) at each time. The solution of this problem for the two limiting cases ($\nu = 0$ and $\nu = 1$) was found numerically on a computer; the Jacobian in (1.38) was determined through a simultaneous calculation of three closely spaced trajectories, i.e., the trajectories of fluid particles having the initial coordinates

$$\begin{aligned} x(0) &= x_0, & y(0) &= y_0; \\ x^*(0) &= x_0 + \delta x_0, & y^*(0) &= y_0; \\ x^{**}(0) &= x_0 & y^{**}(0) &= y_0 + \delta y_0. \end{aligned} \tag{3.33}$$

For small values of δx_0 and δy_0 we have

$$\frac{\rho(x, y)}{\rho_0(x_0, y_0)} = \frac{\delta x_0 \delta y_0}{|(x^* - x)(y^{**} - y) - (x^{**} - x)(y^* - y)|}. \tag{3.34}$$

Before we turn to a discussion of the results of this numerical calculation we will discuss some asymptotic properties of the solution.

A. Asymptotic Properties of the Solution

Let us consider the initial plasma flow far from the current sheet. We assume that the parameter b/z is small. Expanding the complex potential (3.8) in a series in terms of this small parameter, and neglecting the quadratic terms, we find

$$F(z, t) \simeq \frac{\alpha}{2} z^2 - \frac{\alpha}{4} b^2 + \frac{\Gamma}{2\pi} Ln \frac{2z}{b}. \tag{3.35}$$

The vector potential A(x, y, t) is the real part of this expression, given by

$$A(x, y, t) \simeq \frac{\alpha}{2}(x^2 - y^2) - \frac{\alpha}{4} b^2 + \frac{\Gamma}{2\pi} \ln \frac{2\sqrt{x^2 + y^2}}{b}. \tag{3.36}$$

It corresponds to the frozen-in condition (the first integral of motion) in the form

$$x^2 - y^2 - \frac{b^2}{2} + \frac{\Gamma}{2\pi\alpha} \ln \frac{4(x^2 + y^2)}{b^2} = x_0^2 - y_0^2; \tag{3.37}$$

the components of the magnetic field are

$$H_x = -\alpha y + \frac{\Gamma b^2}{2\pi\alpha} \frac{y}{x^2 + y^2}, \tag{3.38}$$

$$H_y = -\alpha x - \frac{\Gamma b^2}{2\pi\alpha} \frac{x}{x^2 + y^2}. \tag{3.49}$$

Since the plasma is initially at rest, the plasma flow at small values of t (and far from the current sheet at any values of t) is slow; i.e., the inequality $\delta \ll 1$ is satisfied. We can therefore replace Eq. (3.1) by the condition that the velocity is orthogonal to the lines of force:

$$\dot{x} A'_y - \dot{y} A'_x = 0. \tag{3.40}$$

Integrating Eq. (3.40), neglecting the small (quadratic) terms in (3.38) and (3.39), we find the second integral of motion to be

$$xy = x_0 y_0. \tag{3.41}$$

According to (1.38) the change in the density of the fluid particle is, with the same accuracy,

$$\rho(x, y, t)/\rho_0 = (x^2+y^2)\left\{(x^2+y^2)^2 - \left(x^2-y^2-\frac{b^2}{4}\right) b^2\left[1+\nu\ln\frac{4(x^2+y^2)}{b^2}\right]\right\}^{-1/2}. \tag{3.42}$$

The density increases (decreases) if

$$A_0 > (<) \frac{\alpha b^2}{4}\left[1+\nu\ln\frac{4(x^2+y^2)}{b^2}\right]. \tag{3.43}$$

In the case $\nu = 0$ this result converts to the result found previously in [16].

B. Results of Numerical Calculation

The problem was solved numerically for the parameter values $k = 1$, $\tau = 0.01$, and $l = 10$ for two limiting cases: $\nu = 0$ and $\nu = 1$. The general case falls between these limits and is geometrically similar to the case $\nu = 0$. Let us examine some results of the numerical calculation.

1. Case of a Vanishing Total Current in the Sheet ($\nu = 0$). Figure 22a shows the trajectories of fluid particles on a line of force $A(x, y, t) = A(x_0, y_0, 0) \equiv A_0 = 0.1$ (the unit for the measurements of A is chosen to be αL^2) and the position of this line of force at subsequent times (see also Figs. 21a and 21b). The dashed curves show the line of force $A_0 = 0$ at the same times. Analogously, Fig. 22b shows the deformation of the line of force $A_0 = -0.1$ and the corresponding particle trajectories.

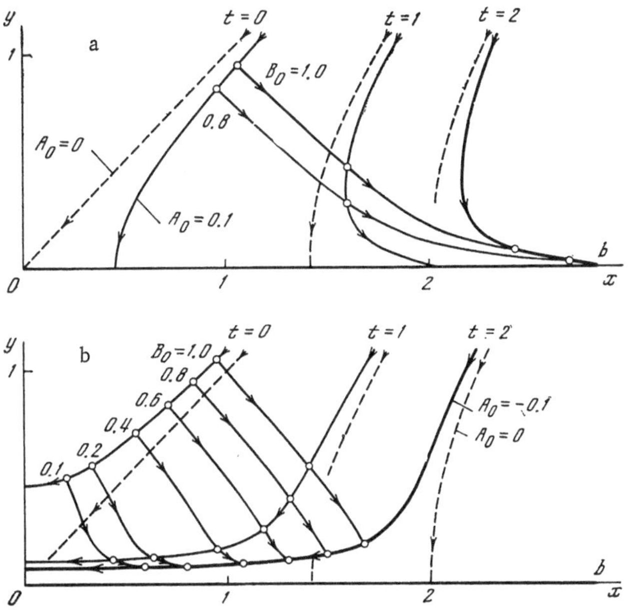

Fig. 22. Motion of fluid particles bound to the lines of force $A_0 = 0.1$ (a) and $A_0 = -0.1$ (b) by the frozen-in condition as the current sheet increases in length from zero to $b \simeq 2.8$ ($t = 2$) in the case $\nu = 0$. Dashed curves) $A_0 = 0$; heavy solid curves) the current sheet at $t = 0$ (along the x axis from 0 to b).

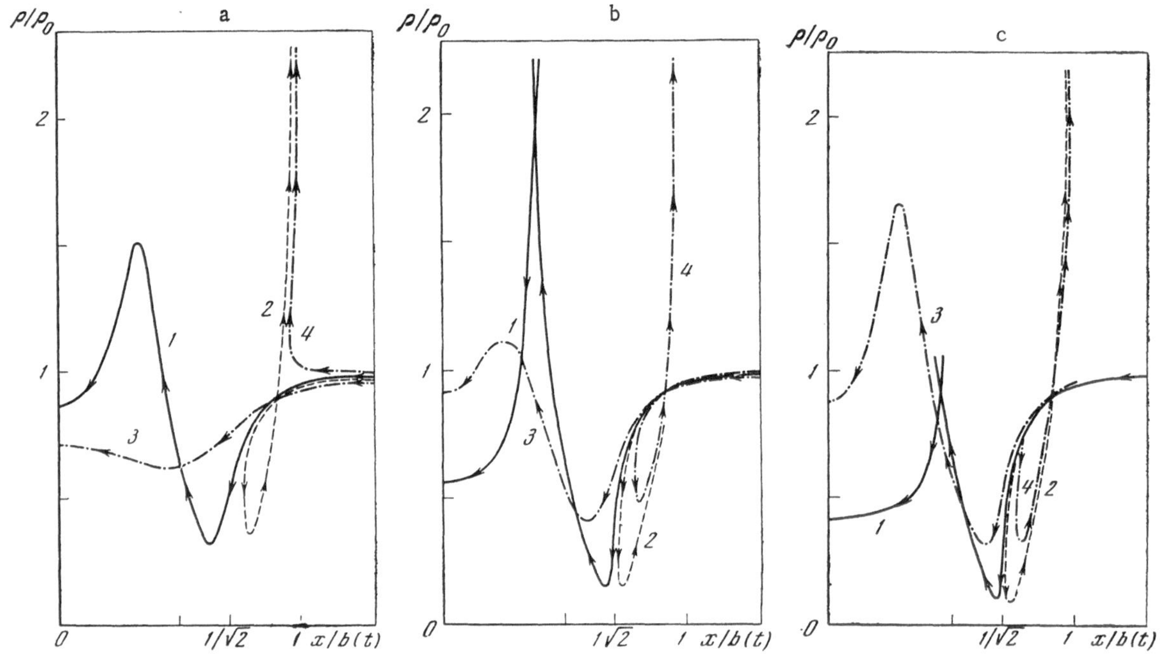

Fig. 23. Profile of the plasma density along the lines of force $A_0 = -0.01$ (1), 0.01 (2), -0.1 (3), and 0.1 (4) as a function of x/b(t) at the times t = 0.2 (a), 1 (b), and 2 (c) in the case $\nu = 0$. The arrows show the directions around the lines of force, as in Fig. 21.

Figure 23 shows the plasma density profiles along the same lines of force ($A_0 = -0.01$, -0.1, and 0.01) as a function of x/b(t) at the times t = 0.2, 1, and 2. These lines of force come very close to the current sheet and run basically parallel to it and to each other. By constructing the plasma-density profile along these lines of force we can graphically demonstrate the high degree of plasma uniformity near a current sheet.

The lines of force $A_0 = -0.1$ and 0.1 are farther from the initial null point than are the lines $A_0 = -0.01$ and 0.01 (Fig. 21a), so that at time t = 0.2 (Fig. 23a) the asymptotic solution characteristic of early times still holds for these former lines of force [see (3.43); specifically, the density decreases along the entire line of force $A_0 = -0.1$, while it increases along the entire line of force $A_0 = 0.1$. On the other hand, on the lines of force $A_0 = -0.01$ and 0.01, which come closer to the original null point and the current sheet which develops from it, the density profile is characteristic of "late" times. For example, the plasma density at the line $A_0 = -0.01$ near the center of the sheet, i.e., at small x/b, decreases. Then, approximately half-way between the center of the current sheet and the null point, which is at the surface of the sheet at, $x/b = 1/\sqrt{2}$, the plasma density increases to a maximum and then falls rapidly as the null point is approached. The same decrease in the plasma density near the null point is found for the line $A_0 = 0.01$. The plasma density near the edge of the current sheet (x/b = 1), like the magnetic field, increases without bound in this approximation of an infinitesimally thin sheet.

With distance from the current sheet ($x/b \gtrsim 1.3$) all the plasma profiles tend toward the original uniform profile with $\rho = \rho_0$.

At t = 1 (Fig. 23b) the density decreases near the center of the sheet and near the null point on all the lines of force mentioned above. Comparison of the densities on different lines

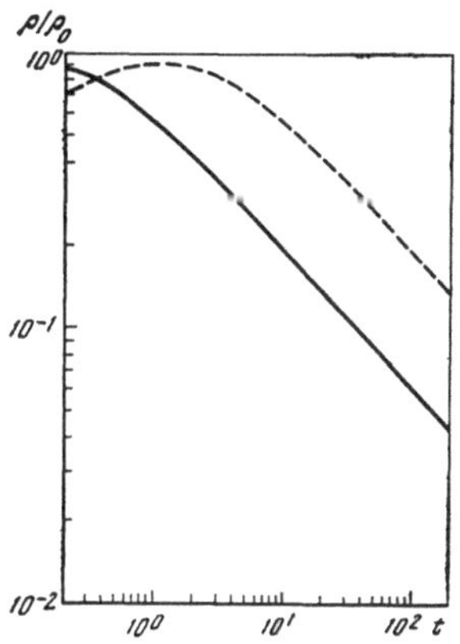

Fig. 24. Change in the density of fluid particles bound to the lines of force $A_0 = -0.01$ (solid curve) and $A_0 = -0.1$ (dashed curve) in motion along the y axis toward the centers of the current sheet.

of force leads to the conclusion that there is a density gradient across the lines of force: The decrease in the plasma density is more pronounced the closer the line of force comes to the center of the current sheet or to the null point at the surface of the current sheet. Comparison with Fig. 23c (t = 2) shows that as the current sheet develops, the plasma density becomes progressively lower in both regions in which the density decreases.

Figure 24 shows the decrease of the plasma density as a function of the time near the center of the developing current sheet at the lines of force $A_0 = -0.01$ and -0.1. At late times the density falls off roughly in proportion to $t^{-1/2}$: the $\rho = \rho(t)$ curves, shown in full logarithmic scale, blend smoothly into straight lines with a slope $-1/2$.

At sufficiently large values of t the motion along the lines of force near the center is described by $x(t) = u_1 t + f(t)$, where $u_1 = \text{const}$ and $|f(t)| \ll u_1 t$.

An interesting kinematic feature of the problem is the existence of regions which could be called "regions of mutual penetration of plasma streams." These regions appear because in this approximation of a strong field and a cold plasma we are neglecting the gas-pressure gradient in the MHD equation of motion. As a result, two fluid particles moving in opposite directions along the force tube do not "sense" each other. This type of motion of fluid particles usually occurs in regions of compression, so that in a numerical integration of the basic system of differential equations the trajectories of the fluid particles formally intersect in compression regions, and the density profile becomes a multivalued function of x/b. This feature can be seen in Figs. 23b and 23c in the plasma-compression region near the current sheet, roughly half-way between the rarefaction regions. Similar regions of mutual penetration of plasma streams occur outside the current sheet on the x axis. There again the plasma density increases.

To analyze these regions of plasma compression we must take into account plasma heating, the various plasma-cooling mechanisms, and (perhaps) other physical processes. These regions are clearly of interest as possible sites for "condensation" of the plasma around the current sheet into dense, hot blots. However, it is not our purpose here to study these compression regions near the sheets; in fact it would be incorrect to pursue this point in the present approximation, as we see from the fact that the trajectories of the fluid particles interesect. In this connection we recall that the physical condition proposed in [19] for the appearance of current sheets ($\mathbf{H} = 0$, $\mathbf{E} \neq 0$) can also be discussed formally in terms of trajectories: There can be no continuous plasma motion (without intersections or discontinuities) for which the hyperbolic magnetic field shown in Fig. 7a takes the form in Fig. 7b. Only through the introduction of a current sheet (Fig. 7c) can the continuity of the trajectories be conserved.

Let us consider the trajectory intersection from this standpoint in two cases: (1) that of a very fast increase of a magnetic dipole; (2) the compression region as plasma moves near a current sheet.

As a dipole moment increases rapidly, discontinuities (oblique shock waves and rotational discontinuities) can appear. However, such discontinuities appear when the typical plasma flow velocity becomes comparable to the Alfvén velocity, i.e., when the initial assumption $\varepsilon^2 \ll 1$ is violated.

In the second example the trajectories intersect in regions where the strong-field approximation is inapplicable – in which the sound velocity becomes comparable to the Alfvén velocity, and the gas-pressure gradient must be taken into account.

Accordingly, the mathematical requirement that the trajectories of the fluid particles be single-valued holds for a broader class of phenomena and can be taken as a quite general condition; it includes, first, the appearance of current sheets upon a violation of the frozen-in equation (the physical condition $\mathbf{H} = 0$, $\mathbf{E} \neq 0$) and, second, the applicability limits of the original assumptions (the assumptions that the parameters γ^2 and ε^2 are small).

In connection with the compression regions near a current layer we would simply add that the influence of the compression regions on the plasma rarefaction regions is felt after a delay proportional to V_S^{-1}; outside the compression region we have $V_S \ll V_A$.

2. Case in Which There Are No Return Currents ($\nu = 1$). We see from Fig. 21c that only the lines of force with $A_0 = -0.1, -0.01, \ldots$ pass close to the current sheet and are essentially parallel to it. For comparison with the previous case $\nu = 0$ we show in Fig. 25 the plasma density profiles along these lines of force at time $t = 5.6$ ($b \simeq 2$).

As in the case $\nu = 0$ there is a density decrease near the null point, which now coincides with the edge of the sheet, and also near the center of the current sheet. The decrease in the plasma density, as in the case $\nu = 0$, is more pronounced the closer the line of force passes to the current sheet.

Accordingly, the results of the numerical calculation confirm that the plasma motion near a current sheet is accompanied by the appearance of regions with a pronounced density decrease. These regions lie near the center of the current sheet and near the null points at its surface. The importance of this conclusion in connection with the problem of particle acceleration in a plasma near magnetic null lines is discussed in [16, 17, 115].

§ 3. Hydrodynamic Plasma Flow near a Steady-State Current Sheet

In contrast with the preceding problem we now assume that the current sheet already exists and has a constant width 2b. A constant and uniform electric field $\mathbf{E}_0$ is directed paral-

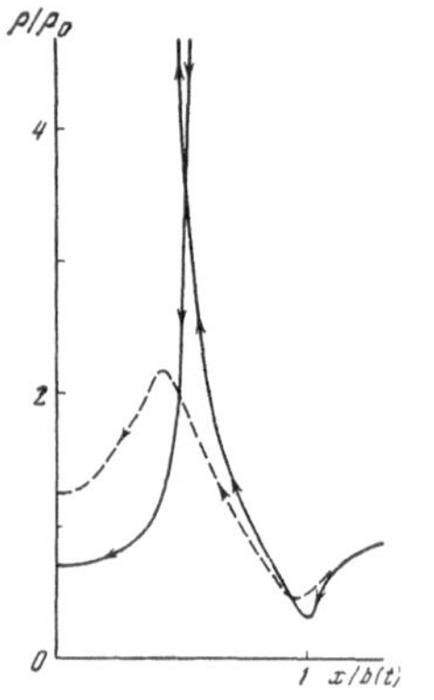

Fig. 25. Profile of the plasma density along the lines of force in the limiting case in which there is no return current, with $\nu = 1$, $t = 5.6$ ($b \simeq 2$) for $A_0 = -0.01$ (solid curve) and $A_0 = -0.1$ (dashed curve).

lel to the current sheet. In this case the vector potential is

$$A(x, y, t) = A(x, y) + A(t). \tag{3.44}$$

The dependence A(t) is determined from Eq. (3.13) with b = const. Transforming to the dimensionless quantities b, t, A(t), and l in this equation, we find

$$\frac{b^2}{4} - \nu \frac{b^2}{2} \ln \frac{b}{2l} - \left(\frac{A^*}{\alpha L^2}\right) A(t) = \left(\frac{cE_0 T}{\alpha L^2}\right) t. \tag{3.45}$$

Here L, T, and A^* are the units of length, time, and magnetic flux. Assuming

$$\frac{A^*}{\alpha L^2} = 1, \qquad \frac{cE_0 T}{\alpha L^2} = 1, \tag{3.46}$$

we find the dimensionless equation

$$\frac{b^2}{4} - \nu \frac{b^2}{2} \ln \frac{b}{2l} - A(t) = t, \tag{3.47}$$

which determines the time dependence of the vector potential:

$$A(t) = A(0) - t, \tag{3.48}$$

where

$$A(0) = \frac{b^2}{4} - \nu \frac{b^2}{2} \ln \frac{b}{2l}. \tag{3.49}$$

As the unit of length we choose the half-width b of the current sheet; then

$$L = b, \qquad A^* = \alpha b^2, \qquad T = \frac{\alpha b^2}{cE_0}. \tag{3.50}$$

It follows from (3.44) and (3.48) that

$$A''_{xt} = A''_{yt} = A''_{tt} = 0, \tag{3.51}$$

and the derivatives A'_x, A'_y, A''_{xx}, and A''_{xy} are given by Eqs. (3.20), (3.21), (3.25), and (3.26), respectively. In contrast with the preceding problem, we now have

$$A'_t = -1 \tag{3.52}$$

everywhere, including at the initial time. Accordingly, at t = 0 we must specify, in addition to the coordinates

$$x(0) = x_0, \qquad y(0) = y_0 \tag{3.53}$$

of the fluid particle, the velocity component perpendicular to the line of force on which the given particle is located. In other words, we must specify

$$u(0) = V_{\perp x}(0) = -\frac{A'_t A'_x}{A'^2_x + A'^2_y}, \tag{3.54}$$

$$v(0) = V_{\perp y}(0) = -\frac{A'_t A'_y}{A'^2_x + A'^2_y}. \tag{3.55}$$

We assume that the velocity component $\mathbf{v}_\parallel$ along the magnetic field is initially zero.

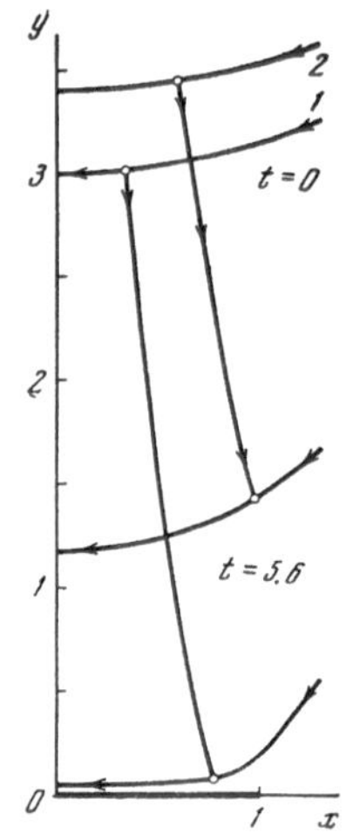

Fig. 26. Trajectories of fluid particles bound to lines of force in the case $\nu = 1$, b = 1. 1) $A_0 = -5.65$; 2) $A_0 = -7.0$.

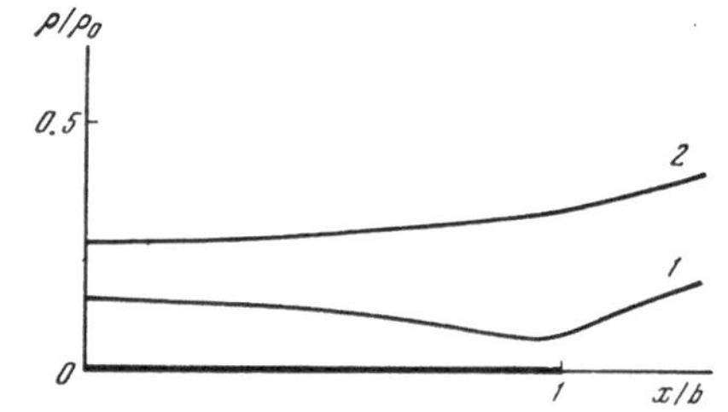

Fig. 27. Profile of the plasma density along the lines of force at time t = 5.6. 1) $A_0 = -5.65$; 2) $A_0 = -7.0$.

This problem, like the preceding problem, was solved numerically; the results for $\nu = 1$ are shown in Figs. 26 and 27.

Figure 26 shows the lines of force $A_0 = -5.65$ $[A_0 \equiv A(x_0, y_0)]$ and $A_0 = -7.0$. At t = 5.6 the first of them approaches the current sheet and is essentially parallel to it. Also shown in this figure are the trajectories of several fluid particles bound to these lines of force.

Figure 27 shows the plasma density profile along these lines of force. As in the case of an existing current sheet, the plasma density decreases near the center of the sheet and the null points.

§ 4. Discussion of Results

These calculations for plasma motion near developing and steady-state current sheets show that in both cases the plasma density decreases at the center of the sheet. To be precise we should speak of a density decrease near the sheet, since the sheet itself and the processes occurring in it lie outside the present treatment.

In the case of a developing sheet the density decreases at the center of the sheet because over the interval in the which the field is applied, of length τ, the particles acquire a velocity along magnetic lines of force in opposite directions on both sides of the y axis. The subsequent inertial continuation of this motion causes the repulsion of plasma from the central region. Here, however, we must take into account two circumstances.

First, the nature of this plasma repulsion can be strongly influenced by the law adopted for the application of the field. We do not rule out the possibility that if this law differs from that adopted in the present calculations [see (3.17)] the plasma flow will be different and the pattern of density changes will also be different. This question requires further study.

Second, in addition to the rarefaction region near the sheet, regions with an elevated plasma density also appear, and on the basis of the results of these calculations we cannot yet conclude that there is an average decrease in the plasma density near the sheet. If there is no such average decrease, the plasma density near a quasisteady sheet will level off as time elapses because of the thermal velocities, and it will be impossible to produce a pronounced rarefaction.

These two comments remain valid for the second problem considered here – that of plasma motion near a current sheet with a steady-state width. In this case the establishment stage was not examined but it was nevertheless shown that the density decreases near the sheet even if the plasma does not initialy spread out along the magnetic field (i.e., with $\mathbf{v}_{\|} = 0$). Furthermore, it follows from Figs. 26 and 27 that the plasma density decreases throughout the region near the sheet, i.e., over the entire length of the magnetic line of force stretching out along the sheet. In contrast with the case of a developing sheet, there are no regions of pronounced increase in the plasma density to which the second of these comments would apply.

We can therefore conclude that for a sheet of steady-state length there is a progressive decrease in density throughout the region near the sheet and that ultimately, if the electric field $E \sim \dot{\beta}$ acts for a sufficiently long time, we reach a situation in which the current sheet is surrounded by a region with an arbitrarily low plasma density. The importance of this conclusion for the the problem of current sheets in plasmas in the laboratory and space is discussed in the first paper in this collection.

Let us briefly summarize the basic results of this paper.

In the first chapter we considered plasma flow in a strong magnetic field having no null points (or lines). In this case the system of equations corresponding to the approximation of a strong field and a cold plasma has no singularities and does have continuous solutions. Plasma flow in a plane dipole magnetic field can serve as an example. In solving this and other two-dimensional problems of MHD it is convenient to treat the (x, y) plane as a complex plane $z = x + iy$ and to associate the function $A(x, y, t)$, which is sinusoidal in the zeroth approximation, with the analytic function $F(z, t) = A(x, y, t) + iB(x, y, t)$. Then it becomes possible to make effective use of the methods of the theory of function of a complex variable to solve two-dimensional problems.

As an example we derived the analytic solution of the linearized problem corresponding to small changes in a plane dipole magnetic field. This problem demonstrates a tendency in the behavior of a plasma in a strong dipole field: As the dipole moment increases, the plasma density at the dipole axis increases in proportion to the magnetic moment, while that in the equatorial plane decreases. This "magnetic raking" and plasma compression are interesting in connection with many applications in astrophysics and the physics of laboratory plasmas. The three-dimensional linearized problem analogous to this problem was solved in [18]. For concrete examples it is necessary to have exact nonlinear solutions, corresponding to arbitrary (not small) changes in a three-dimensional dipole magnetic field. Such solutions have been obtained [28] and applied to two astrophysical problems – the interaction of the interstellar medium with the magnetic field of an expanding envelope and the formation of solar condensations and prominences [71, 54].

It was shown in the first of these applications that the raking of the interstellar plasma by the magnetic field of the expanding envelope of a nova can lead to the appearance of condensations of the interstellar plasma ahead of the envelope. In the second application the magnetic raking of plasma is discussed as a possible mechanism for the formation of certain types of prominences and condensations in the solar chromosphere and corona during local intensifications of the photospheric magnetic field.

The presence of magnetic null lines changes the situation radically. If an electric field is applied along a null line, continuous plasma flow near such a line becomes possible, and a current sheet with the corresponding discontinuity of the magnetic field appears. General conditions for the appearance of current sheets were studied in [19]. Of interest for concrete applications are the problems of the appearance of current sheets during the motion of a plasma in a dipole field.

The second chapter dealt with two such problems – the elongation of the magnetic field of a plane dipole by a plasma flow [87] and the compression of a cylinder in a homogeneously magnetized medium.

In the first problem it was assumed that the field of a plane magnetic dipole is initially frozen in a plasma. The region is bounded by a homogeneously expanding cylindrical surface, at which magnetic flux is conserved. It was shown that as the radius of the boundary increases by a factor of two, a null point appears at the boundary, and a current sheet develops there during the subsequent expansion. A general solution was found for the magnetic field configuration, and the asymptotic picture of the lines of force was found for an infinite increase in the radius of the boundary.

Three applications of the first problem were pointed out.

One deals with the earth's magnetospheric tail. In this tail there is known to be a neutral sheet, separating magnetic fields opposite in direction. The reasons for the appearance of this tail have not yet been finally resolved; according to [98, 99], even the partial penetration of the geomagnetic field into the plasma of the solar wind is sufficient for the appearance of a current sheet with the corresponding magnetic field configuration.

A second field of application can be in solar rays or streamers. The appearance of coronal rays and the appearance of the magnetospheric tail are formally completely analogous. In both cases there is an elongation of a dipole magnetic field by the solar wind; in the corona it is the dipole magnetic field of an extended active region, while in the magnetosphere it is the earth's magnetic field.

A third possible application of the problem is in laboratory experiments involving the flow of a plasma around a plane dipole.

A second problem – that of the magnetic field of a plasma cylinder contracting in a homogeneously magnetized plasma – is a problem of cosmic electrodynamics: the gravitational contraction of a plasma cloud linked by a magnetic field with the external medium. The solution shows that, as a contracting cylinder or sphere is formed from a homogeneously magnetized plasma, the plasma motion leads to the formation of a current sheet. In the case of a sphere the current sheet is a thin equatorial ring, while in the case of a cylinder it consists of two current sheets, requiring a closing contour at infinity.

The third chapter dealt with hydrodynamic plasma flow near a current sheet.

Much progress has now been achieved in our understanding of the fundamental role of magnetic null points (or lines) in plasma dynamics, particularly in space applications. The general thrust of this research has now shifted toward the analysis of the concrete physical problems which arise in a plasma near such points.

Syrovatskii [19] has studied the structure of a strong magnetic field frozen in a plasma in the case of plane two-dimensional flow. He found that a necessary and sufficient condition for the appearance of a current sheet is that there be an electric field parallel to the neutral line. In the third chapter we used these results to continue the analysis of plasma flow in the vicinity of the simplest case of a null line. We determined the trajectories of the fluid (plasma) particles and the behavior of the plasma density near a developing null line and a steady-state current sheet. The basic result of this analysis was the proof of an assertion which has important applications: As a plasma moves near a null line, regions of a pronounced decrease in the plasma density appear. This conclusion is interesting in connection with the general problem of particle acceleration in plasmas near magnetic null lines.

Literature Cited

1. L. D. Landau and E. M. Lifshits (Lifshitz), Electrodynamics of Continuous Media, Addison-Wesley, Reading, Mass. (1962).
2. S. I. Syrovatskii, Usp. Fiz. Nauk, 62:247 (1957).
3. T. Cowling, Magnetohydrodynamics, Interscience, New York (1957).
4. J. A. Shercliff, A. Textbook of Magnetohydrodynamics, Pergamon, New York (1965).
5. J. J. Monoghan, Monthly Not. Roy. Astron. Soc., 132:1 (1966); 134:275 (1966).
6. T. C. Chaim and J. J. Monoghan, Monthly Not. Roy. Astron. Soc., 155:153 (1971).
7. I. W. Roxburg, Monthly Not. Roy. Astron. Soc., 126:67 (1963); 132:347 (1966); 135:329 (1967).
8. L. M. Ozernoi and V. V. Usov, Astrophys. Space Sci., 13:3 (1971).
9. S. Lundquist, Arkir Fys., 5:297 (1952).
10. E. N. Parker, Annual Review of Astronomy and Astrophysics, Vol. 8 (1970), p. 1.
11. N. O. Weiss, Quart. J. Roy. Astron. Soc., 12:432 (1971).
12. E. N. Parker, Astrophys. J., 157:1119 (1969); 162:665 (1970).
13. S. I. Vainshtein and Ya. B. Zel'dovich, Usp. Fiz. Nauk, 106:431 (1972).
14. J. H. Piddington, Monthly Not. Roy. Astron. Soc., 133:163 (1966).
15. L. M. Ozernoi and B. V. Somov, Astrophys. Space Sci., 11:244 (1971).
16. S. I. Syrovatskii, Astron. Zh., 43:340 (1966).
17. S. I. Syrovatskii, Zh. Éksp. Teor. Fiz., 50:1133 (1966).
18. S. I. Syrovatskii, Astrophys. Space Sci., 4:240 (1969).
19. S. I. Syrovatskii, Zh. Éksp. Teor. Fiz., 60:1727 (1971).
20. N. F. Ness, Rev. Geophys., 7:97 (1969).
21. E. N. Parker, in: Physics of the Magnetosphere (ed. R. L. Carovillano, J. F. McClay, and H. R. Radoski), Dordrecht, Holland (1968), p. 3.
22. K. H. Schatten, Cosmic Electrodynam., 2:232 (1971).
23. L. F. Burlaga, "The solar envelope," Preprint X-692-71-400, Goddard Space Flight Center, Greenbelt, Maryland, 1971.
24. I. M. Podgornyi and R. Z. Sagdeev, Usp. Fiz. Nauk, 98:409 (1969).
25. I. M. Podgornyi, É. M. Dubinin, and G. G. Managadze, Preprint D-15, Institute for Cosmic Studies, Academy of Sciences of the USSR, 1970.
26. I. M. Podgornyi and É. M. Dubinin, Cosmic Electrodynam., 2:445 (1972).
27. M. A. Lavrent'ev and B. V. Shabat, Methods of the Theory of Functions of a Complex Variable [in Russian], Gostekhizdat, Moscow (1951).
28. B. V. Somov and S. I. Syrovatskii, Zh. Éksp. Teor. Fiz., 61:621 (1971).
29. É. R. Mustel', Astron. Zh., 33:182 (1956).
30. E. R. Mustel' (Mustel) and A. A. Boyarchuk, Astrophys. Space Sci., 6, 183 (1970).
31. L. A. Artsimovich, Controlled Thermonuclear Reactions [in Russian], Fizmatgiz, Moscow (1961), §§ 8, 20.
32. B. A. Vorontsov-Vel'yaminov, Gaseous Nebulae and Novae [in Russian], Izd. Akad. Nauk SSSR, Moscow (1948).
33. É. R. Mustel', Astron. Zh., No. 534 (1969); No. 544 (1970).
34. G. A. Gurzadyan, Planetary Nebulae [in Russian], Fizmatgiz, Moscow (1962).
35. G. S. Hromov and L. Kohoutek, Catalogue of Galactic Planetary Nebulae, Akademia, Prague (1967).

36. G. S. Hromov and L. Kohoutek, BAC, 19(1):81, 90 (1968).
37. G. P. Kuiper, Astrophys. J., 86:102 (1937).
38. G. P. Kuiper, Publ. Astron. Soc. Pacific, 47:228, 267 (1935); 53:330 (1941).
39. R. G. Aitken, Publ. Astron. Soc. Pacific, 48:340 (1936).
40. Voûte, Publ. Astron. Soc. Pacific, 47:270 (1935).
41. G. Van Biesbroeck, Astrophys. J., 82:433 (1935).
42. G. Van Biesbroeck, Observatory, 62:236 (1939).
43. W. Baade, Publ. Astron. Soc. Pacific, 52:386 (1940); 54:244 (1942).
44. P. W. Merrill, Astrophys. J., 82:413 (1935).
45. M. L. Humason, Publ. Astron. Soc. Pacific, 52:389 (1940).
46. P. Swings and O. Struve, Astrophys. J., 92:295 (1940).
47. I. M. Kopylov, Izv. KrAO, 10:200 (1953).
48. W. Baade, Sky and Telescope, 12:12 (1952).
49. H. C. Van de Hulst, Proceedings of the Symposium on Problems of Cosmical Aerodynamics, Dayton, Ohio, 1951, p. 116.
50. D. B. McLaughlin, Stellar Atmospheres (ed. J. L. Greenstein) (1960), p. 585.
51. E. Schatzman, Ann. Astrophys., 21:1 (1958).
52. S. Sakashita, Astrophys. Space Sci., 14:431 (1971).
53. I. G. Kolesnik, Izv. GAO, 4:63 (1962).
54. B. V. Somov, "Effect of strong magnetic fields on the structure of the solar corona," Proceedings of the 16th Conference of the Moscow Physicotechnical Institute [in Russian], Moscow (1972), p. 68.
55. G. W. Preston and K. Stepien, Astrophys. J., 154:971 (1968).
56. É. S. Brodskaya, Astron. Zh., 47:662 (1970).
57. C. Blanco and F. A. Catalano, Astrophys. J., 75:53 (1970).
58. H. Zirin, in: Mass Motion in Solar Flares and Related Phenomena (ed. Y. Öhman), Stockholm (1968), p. 131.
59. M. McCabe and R. R. Fisher, Solar Phys., 14:212 (1970).
60. E. F. Shaposhnikova, Izv. KrAO, 18:151 (1958).
61. V. Becker, Z. Astrophys., 48:189 (1959).
62. T. Fortini and M. Torelly, in: Mass Motion in Solar Flares and Related Phenomena (ed. Y. Öhman), Stockholm (1968), p. 163.
63. K. D. Kiepenheuer, in: Mass Motion in Solar Flares and Related Phenomena (ed. Y. Öhman), Stockholm (1968), p. 123.
64. A. N. Koval', Izv. KrAO, 33:138 (1965).
65. D. Rust, in: Structure and Development of Solar Active Regions (ed. K. O. Kiepenheuer), IAU Symposium N. 35, Riedel, Dordrecht (1968), p. 77.
66. M. D. Altschuler, C. G. Lilliequist, and Y. Nakagawa, Solar Phys., 5:366 (1968).
67. J. T. Jefferies and F. Q. Orrall, Astrophys. J., 133:946 (1961).
68. R. G. Giovanelli and M. K. McCabe, Austral. J. Phys., 11:191 (1958).
69. A. Schlüter, in: Radio Astronomy (ed. H. C. van de Hulst), IAU Symposium N 4, Cambridge Univ. Press (1957), p. 356.
70. Yu. V. Platov, Solar Phys., 28:477 (1973).
71. Yu. V. Platov, B. V. Somov, and S. I. Syrovatskii, Solar Phys., 30:139 (1973).
72. S. I. Syrovatskii, Acta Phys. Hung., 29:17 Suppl. L (1970).
73. S. I. Syrovatskii, Proceedings of the All-Union Seminar on Interplanetary Research, Leningrad, 1969 [in Russian], p. 7.
74. S. I. Syrovatskii, Proceedings of the International Seminar on the Acceleration of Cosmic Rays at the Sun, Leningrad, 1970 [in Russian], p. 15.
75. P. A. Sweet, Annual Review of Astronomy and Astrophysics, Vol. 7 (1969), p. 149.
76. N. F. Ness, J. Geophys. Res., 70:2989 (1963).

77. A. J. Dessler, in: Physics of the Magnetosphere (R. L. Carovillano, J. F. McClay, and H. R. Radoski, eds.), Dordrecht (1968), p. 65.
78. V. P. Shabanskii (Shabansky), Space Sci. Rev., 12:314 (1971).
79. I. V. Kovalevskii (J. V. Kovalevsky), Space Sci. Rev., 12:187 (1971).
80. S. K. Vsekhsvyatskii, G. M. Nikol'skii, V. I. Ivanchuk, A. T. Nesmenovich, E. A. Ponomarev, G. A. Rubo, and V. I. Cherednichenko, The Solar Corona and Corpuscular Emission in Interplanetary Space [in Russian], Izd. Kievsk. Gos. Un-ta (1965).
81. J. D. Bohlin, Solar Phys., 12:240 (1970); 13:153 (1970).
82. K. H. Schatten, J. W. Wilcox, and H. F. Ness, Solar Phys., 6:442 (1969).
83. M. D. Altschuler and G. Newkirk, Solar Phys., 9:131 (1969).
84. G. Newkirk and M. D. Altschuler, Solar Phys., 13:131 (1970).
85. L. P. Van Speybroeck, A. S. Krieger, and G. S. Vaiana, Nature, 227:818 (1970).
86. J. M. Wilcox, Space Sci. Rev., 8:258 (1968).
87. B. V. Somov and S. I. Syrovatskii, Zh. Éksp. Teor. Fiz., 61:1864 (1971).
88. G. W. Pneuman, Solar Phys., 3:578 (1968); 6:255 (1969).
89. G. W. Pneuman and R. A. Kopp, Solar Phys., 13:176 (1970).
90. S. I. Syrovatskii, Zh. Éksp. Teor. Fiz., 24:622 (1953).
91. I. M. Gordon, Astron. Zh., 45:1002 (1968).
92. I. M. Gordon, V. A. Liperovskii, and V. N. Tsytovich, Preprint No. 35, Institute for Cosmic Studies, Academy of Sciences of the USSR, 1970.
93. F. Axisa, Y. Avignon, M. J. Martres, M. Pick, and P. Simon, Solar Phys., 19:110 (1971).
94. V. N. Zhugulev and E. A. Romishevskii, Dokl. Akad. Nauk SSSR, 126:521 (1959); 127:1001 (1959).
95. T. Unti and G. Atkinson, J. Geophys. Res., 73:7319 (1968).
96. J. D. Coll and J. H. Huth, Phys. Fluids, 2:624 (1959).
97. T. Obayashi, J. Geophys. Rev., 69:861 (1964).
98. S. I. Syrovatskii, Research on Geomagnetism, Aeronomy, and Solar Physics [in Russian], Nauka, Moscow (1972), No. 23, p. 195.
99. B. V. Somov and S. I. Syrovatskii, Proceedings of the International Seminar on Particle Acceleration in Space, Leningrad, 1971 [in Russian], p. 272.
100. S. I. Syrovatskii, Critical Problems of Magnetospheric Physics, Proceedings of the COSPAR Symposium, 1972, IUSTP, Washington (1972), p. 35.
101. P. Obertz, Geomagnetizm i Aéronomiya, 12:280 (1972); 13:896 (1973).
102. S. M. Hamberger and M. Friedman, Phys. Rev. Letters, 21:874 (1968).
103. L. Mestel, Monthly Not. Roy. Astron. Soc., 119:223 (1959).
104. L. Mestel, Monthly Not. Roy. Astron. Soc., 133:265 (1966).
105. L. Mestel and P. A. Strittmatter, Monthly Not. Roy. Astron. Soc., 137:95 (1967).
106. Yu. D. Zhugzhda, Geomagnetizm i Aéronomiya, 6:506 (1966).
107. K. L. McDonald, Amer. J. Phys., 22:586 (1954).
108. P. A. Sweet, Nuovo Cimento, 8:188, Suppl. X (1958).
109. P. A. Sweet, in: Stellar and Solar Magnetic Fields (ed. R. Lüst), Amsterdam (1965), p. 377.
110. S. Chapman and P. C. Kendall, Proc. Roy. Soc., A271:435 (1963).
111. V. S. Imshennik and S. I. Syrovatskii, Zh. Éksp. Teor. Fiz., 59:990 (1967).
112. S. I. Syrovatskii, Zh. Éksp. Teor. Fiz., 54:1422 (1968).
113. B. Coppi and A. B. Friedland, Astrophys. J., 169:369 (1971).
114. S. I. Syrovatskii and B. V. Somov, Proceedings of the International Seminar on Particle Acceleration in Space, Leningrad, 1971 [in Russian], p. 106.
115. S. I. Syrovatskii, Tr. FIAN, 74:3 (1974).

NUMERICAL INTEGRATION OF THE MHD EQUATIONS NEAR A MAGNETIC NULL LINE

N. I. Gerlakh and S. I. Syrovatskii

The MHD equations are integrated numerically for a homogeneous plasma initially at rest near a null line in a hyperbolic magnetic field. The finite plasma conductivity is taken into account. A solution of the linearized equations far from the null line – a cylindrical potential-perturbation wave converging on the null line – is adopted as a boundary condition. This perturbation arises when a constant external electric field is applied parallel to the null line. As the null line is approached, the wave becomes nonlinear and leads to the appearance of a thin current sheet. There is no evidence of any kind for the system of slow-mode shock waves corresponding to the Petschek model. Within the current sheet the plasma density increases, while the density just outside the sheet decreases. The plasma flows along the sheet at approximately the Alfvén velocity. A return current flows at the edges of the sheet.

Magnetic null lines have attracted interest for many years as possible sites of a rapid dissipation of magnetic energy in a plasma, primarily in connection with the problem of solar flares [1-7]. Syrovatskii [4, 5] found certain unusual features in the plasma flow near null lines: a rapid increase in the current density and the appearance of a current sheet, on the one hand, and a decrease in the plasma density, on the other.

These effects have consequences which are important for applications and which are discussed in [4-7].

Since the conclusions reached by Syrovatskii in [4, 5] were actually based on a study of small perturbations, the precise behavior of the plasma and the magnetic field in the highly nonlinear region near a null line has remained uncertain. An exact particular solution of the problem found by Imshennik and Syrovatskii [8] (an analogous solution for the case of an incompressible medium was found by Chapman and Kendall [9]) verified Syrovatskii's conclusion of a progressive increase in the current density but did not verify the decrease in plasma density found in [4]. The reason may be that the particular solution of [8] requires special initial and boundary conditions [10] – different from those in the formulation of the problem in [4].

It was therefore necessary to pursue the study in the nonlinear (and most interesting) stage, for which a general analytic solution cannot be found.

Below we give a formulation of the problem and report a numerical integration of the MHD equations for two-dimensional plasma flow near the null line of a strong magnetic field.

We assume the plasma to be initially at rest, and we assume that the magnetic field is a potential field. As a boundary condition we adopt a solution of the linearized equations which follows from the physical formulation of the problem; this solution, which is valid far from the null line, consists of a cylindrical wave converging on the null line [5].

The calculations show that a thin current sheet arises near the null line and separates regions in which the magnetic fields are opposite in direction. Just outside the sheet the calculations reveal a decrease of the plasma density from its initial value, but this effect is small in the initial stage of the process for which the calculations were carried out. Further study will be required to determined whether a pronounced density decrease occurs near the sheet. An important result of these calculations is that the system of standing slow shock waves proposed by Petschek [3] does not arise near the null line with the boundary conditions used here. An analogous conclusion was reached by Stevenson [11] on the basis of a numerical integration of the MHD equations for other initial and boundary conditions.

1. Equations for Two-Dimensional Flow

We consider two-dimensional plane flows in which all properties are independent of the coordinate z and in which the velocity **v** and the magnetic field **H** are parallel to the z = 0 plane. Then the field **H** and the current density **j** can be expressed in terms of a vector potential **A** which has only a z component, $A = A_z$:

$$\mathbf{H} = \operatorname{rot} \mathbf{A} = \left\{ \frac{\partial A}{\partial y}, \ -\frac{\partial A}{\partial x}, \ 0 \right\}, \qquad \mathbf{j} = \frac{c}{4\pi} \operatorname{rot} \mathbf{H} = \frac{c}{4\pi} \{0, \ 0, \ -\Delta A\}. \tag{1}$$

We note that the equation

$$A(x, y, t) = \text{const} \tag{2}$$

defines a family of magnetic lines of force for a fixed time t.

Taking into account the finite conductivity σ of the medium, we can write the MHD equations for plane two-dimensional flow as

$$\frac{\partial A}{\partial t} = \mathbf{v}\nabla A + \nu_m \Delta A, \qquad \frac{\partial \rho}{\partial t} = -\operatorname{div} \rho\mathbf{v},$$
$$\frac{\partial \mathbf{v}}{\partial t} = -(\mathbf{v}\nabla)\mathbf{v} - \frac{\nabla p}{\rho} - \frac{1}{4\pi\rho}\Delta A \nabla A, \quad p = p_0\left(\frac{\rho}{\rho_0}\right)^{\gamma}. \tag{3}$$

Here ρ and p are the plasma density and pressure, and $\nu_m = c^2/4\pi\sigma$ is the magnetic viscosity. We assume that the conductivity is high, there is no heat transfer, and the flow is correspondingly isentropic and has a ratio of specific heats $\gamma = 5/3$.

As the units of length, time, velocity, vector potential, density, pressure, and current density we adopt

$$r_0, \ t_0 = r_0/V_A, \quad V_A = H_0/\sqrt{4\pi\rho_0}, \quad H_0 r_0, \ \rho_0, \ p_0, \text{ and } cH_0/4\pi r_0, \tag{4}$$

where H_0 is a typical value of the magnetic field. Then we can rewrite Eqs. (3) in the dimensionless form

$$\frac{\partial A}{\partial t} = -\mathbf{v}\nabla A + \nu \Delta A,$$
$$\frac{\partial \rho}{\partial t} = -\operatorname{div} \rho\mathbf{v},$$

$$\frac{\partial \mathbf{v}}{\partial t} = -(\mathbf{v}\nabla)\,\mathbf{v} - r_s^2 \frac{\nabla p}{\rho} - \frac{\Delta A}{\rho} \nabla A,$$

$$p = \rho^\gamma \quad (5)$$

Now these equations depend on the two dimensionless parameters

$$r_s^2 = \frac{4\pi p_0}{H_0^2} = \frac{s^2}{\gamma V_A^2} \quad \text{and} \quad \nu = \frac{\nu_m}{r_0 V_A} = \frac{c^2}{\sigma r_0 H_0} \sqrt{\frac{\rho_0}{4\pi}}, \quad (6)$$

where $s^2 = \gamma p_0/\rho_0$ is the square of the sound velocity. The magnetic Reynolds number is evidently ν^{-1}.

We assume that the plasma is initially at equilibrium ($v = 0$) with a uniform density ρ_0 and pressure p_0 and with a hyperbolic potential magnetic field:

$$A(r, \varphi, 0) = \frac{h_0}{2} r^2 \cos 2\varphi. \quad (7)$$

Here and below we use the cylindrical coordinates r, φ, z. According to (7) the magnitude of the original magnetic field is $H(r, \varphi, 0) = h_0 r$, where h_0 is the initial field gradient near the null line. Correspondingly, we assume $H_0 = h_0 r_0$ in Eqs. (4) and (6).

Then the initial conditions (for $t = 0$) take the dimensionless form

$$\mathbf{v}(r, \varphi, 0) = 0, \quad \rho(r, \varphi, 0) = 1, \quad p(r, \varphi, 0) = 1,$$

$$A(r, \varphi, 0) = \frac{1}{2} r^2 \cos 2\varphi. \quad (8)$$

As the boundary condition at the circle $r = 1$ we adopt a cylindrically symmetric potential perturbation [4, 5]

$$A(1, \varphi, t) = \frac{1}{2} \cos 2\varphi - a(t), \quad (9)$$

where $a(t)$ is a given function of the time [$a(0) = 0$]. The specification of boundary condition (9) is not sufficient for solving system (5), which is equivalent to a homogeneous symmetric hyperbolic system of first-order equations, for the functions H_r, H_φ, v_r, v_φ, and ρ; five boundary conditions may be necessary to find the general solution of such a system of equations. Below we adopt the formulation of the problem discussed in [4, 5] in which we have

$$r_s^2 \ll 1 \quad \text{and} \quad \nu \ll 1 \quad (10)$$

and in which the solution far from the null line is a small-amplitude converging cylindrical wave.

2. Small-Amplitude Waves

For small perturbations of the original equilibrium state in (7) we find, after linearizing Eqs. (5) and using conditions (10), the following equation for the potential perturbation $A'(t, r, \varphi)$ [4, 5]:

$$\frac{\partial^2 A'}{\partial t^2} = r \frac{\partial}{\partial r}\left(r \frac{\partial A'}{\partial r}\right) + \frac{\partial^2 A'}{\partial \varphi^2}. \quad (11)$$

The solution of this equation corresponding to a cylindrically symmetric converging wave is

$$A'(t, r, \varphi) = a(t + \ln r), \quad (12)$$

where $a(\tau)$ is an arbitrary function satisfying the condition $a(0) = 0$ [cf. (9)]. Also using the other equations in the linearized system (5) we find

$$\begin{aligned}
A(r, \varphi, t) &= \frac{1}{2} r^2 \cos 2\varphi - a(\ln r + t), \\
v_r(r, \varphi, t) &= \frac{\dot a}{r} \cos 2\varphi, \\
v_\varphi(r, \varphi, t) &= -\frac{\dot a}{r} \sin 2\varphi, \\
\rho(r, \varphi, t) &= 1 + \frac{2\cos 2\varphi}{r^2}\left(a - \frac{\dot a}{2}\right), \\
H_r(r, \varphi, t) &= -r \sin 2\varphi, \\
H_\varphi(r, \varphi, t) &= -r \cos 2\varphi + \frac{\dot a}{r}, \\
j &= \frac{\ddot a}{r^2}.
\end{aligned} \qquad (13)$$

Here $\dot a(\ln r + t) = da/dt$.

For the region under consideration here, $r_s \ll r \leq 1$, solution (12), (13) holds only if

$$\begin{aligned}
a(\ln r + t) &\ll r^2/2, \\
\dot a(\ln r + t) &\ll r^2.
\end{aligned} \qquad (14)$$

Below we assume that these conditions are satisfied at the boundary of the region, $r = 1$, for the entire time interval under consideration, $0 \leq t \leq t_c$, where t_c is bounded by the conditions

$$a(t_c) \ll \frac{1}{2}, \qquad \dot a(t_c) \ll 1. \qquad (15)$$

We use the values of the functions in (13) at $r = 1$ as boundary conditions for the rigorous nonlinear problem defined by Eqs. (5) and (8) in the region $r \leq 1$. Here we should bear in mind that the existence of a region in which hyperbolic equations (5) are not independent makes the specification of all the functions in (13) at the $r = 1$ boundary superfluous and even contradictory. They are contradictory if the linearized solution in (13) breaks down because the boundary of a strong reflected wave is approached at some time t_p. It may turn out that $t_p < t_c$, so that the duration of the calculation must be monitored in each specific case in this formulation of the problem.

3. Numerical Integration

System (5) was solved by a numerical integration on a computer. Exploiting the symmetry of the problem we considered only the following ranges of the cylindrical coordinates (r, φ):

$$\begin{aligned}
0 &\leqslant r \leqslant 1, \\
0 &\leqslant \varphi \leqslant \pi/2.
\end{aligned} \qquad (16)$$

The derivatives appearing in the equations are approximated on the basis of a local single-layer difference scheme, i.e., a scheme in which it is possible to calculate any function at each calculation point on the basis of the preceding material alone. At points of one kind (indicated by the crosses) the velocity components along r and φ, i.e., $u = v_r$ and $v = v_\varphi$, are calculated. At points of the other kind (circles) the density ρ and the vector potential A are calculated. The points of one kind are shifted in time half a step with respect to the points of the other kind. Figure 1 shows a projection of the grid onto the (r, φ) plane. For calculations at

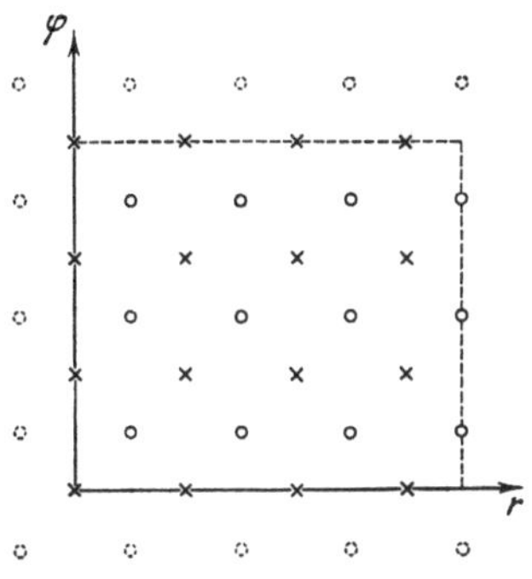

Fig. 1. Calculated grid. Crosses) for values of v_r and v_φ; circles) A and ρ.

a particular point the four nearest points of the other kind and four points of the same kind are used. All the coefficients are averaged over these points, and the derivative are calculated.

The following boundary conditions are specified:

$$\frac{\partial v_r}{\partial \varphi}=0, \quad v_\varphi=0, \quad \frac{\partial \rho}{\partial \varphi}=0, \quad \frac{\partial A}{\partial \varphi}=0 \tag{17}$$

at $\varphi = 0$ and $\varphi = \pi/2$; and

$$v_r=0, \quad v_\varphi=0, \quad \frac{\partial \rho}{\partial r}=0 \tag{18}$$

at r = 0. At r = 1, conditions (13) are used. In accordance with these conditions it is convenient to assume that crosses make up the $\varphi = 0$, $\varphi = \pi/2$, and r = 0 boundaries and that circles make up the right-hand boundary (r = 1). For convenience in writing the boundary conditions at the left, the top, and the bottom (Fig. 1) we introduce fictitious points half a spacing outside the region.

The values in (8) are adopted as the initial values. The time step is chosen to retain stability, i.e., $\max(v + s)\,\Delta t \leq \min(\Delta r, \Delta\varphi)$, where $s^2 = \gamma p/\rho$. The perturbation at the right-hand boundary of the region [see (9)] is assumed a linear function of the time:

$$a(t)=\varkappa t. \tag{19}$$

As was mentioned previously, the use of conditions (13) at the right-hand boundary is justified only for a restricted time interval. As a check, calculations were carried out for versions of the problem in which only some of the conditions in (13) were specified at the right-hand boundary – those necessary for taking into account the region in which Eqs. (5) are not independent. The other properties were determined from the relations imposed on the characteristics. Comparison of the results showed that boundary condition (13) can be used up to times corresponding to values $a(t) \simeq 0.5$–0.7 in (19).

4. Calculated Results and Discussion

Calculations were carried out for two classes of problems; in one the medium was assumed ideally conducting ($\nu = 0$), while in the other the finite conductivity was taken into account. Because of conditions (15) and (19) this boundary-value problem is meaningful only if $\varkappa \ll 1$. For the calculations we adopted the values

$$\varkappa=0.003, \quad 0.03, \quad 0.1, \quad \text{and} \quad 0.3. \tag{20}$$

Since we are interested in the strong-field case [see (10)], we selected the values

$$r_*=0.01, \quad 0.1, \quad \text{and} \quad 0.3. \tag{21}$$

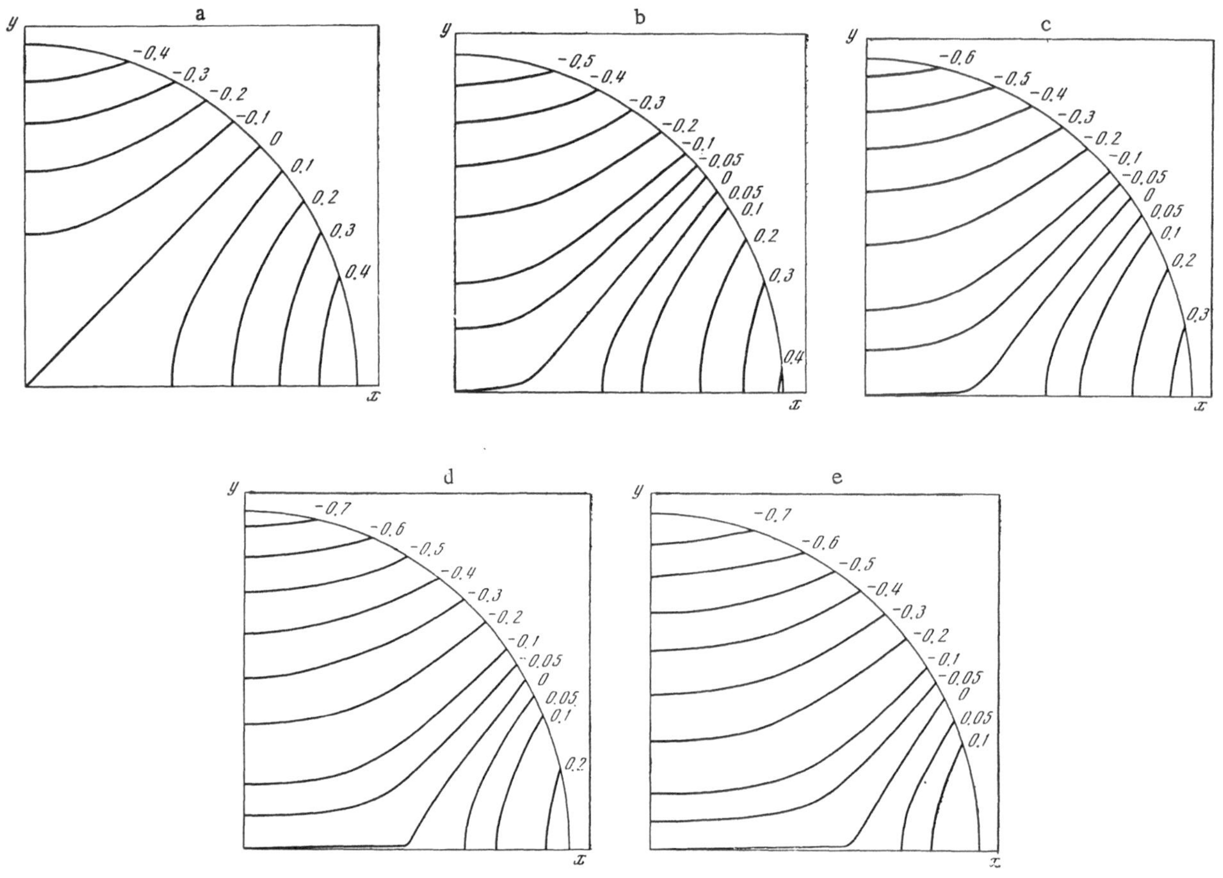

Fig. 2. Magnetic lines of force at various times. a) Initial pattern $t = 0$, $A_0 = (r^2/2)\cos 2\varphi$; b-e) $A(r, \varphi, t) = \text{const}$ for $\nu = 0$, $\varkappa = 0.03$, $r_s = 0.1$; b) $t = 3$; c) 5; d) 8; e) 10.

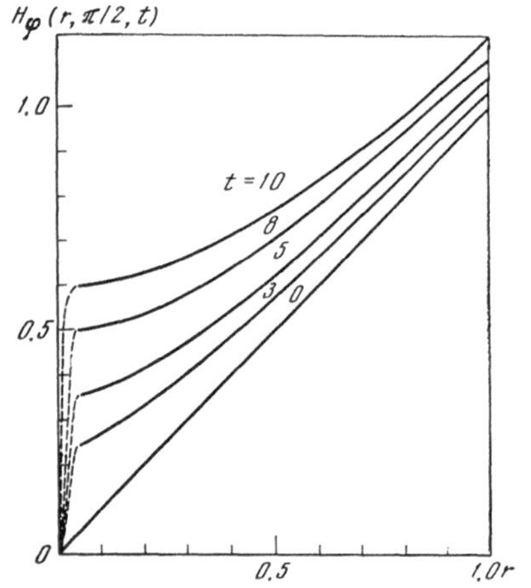

Fig. 3. Horizontal component of the magnetic field $H_\varphi(r, \pi/2, t)$ for $\nu = 0$, $\varkappa = 0.03$, and $r_s = 0.1$.

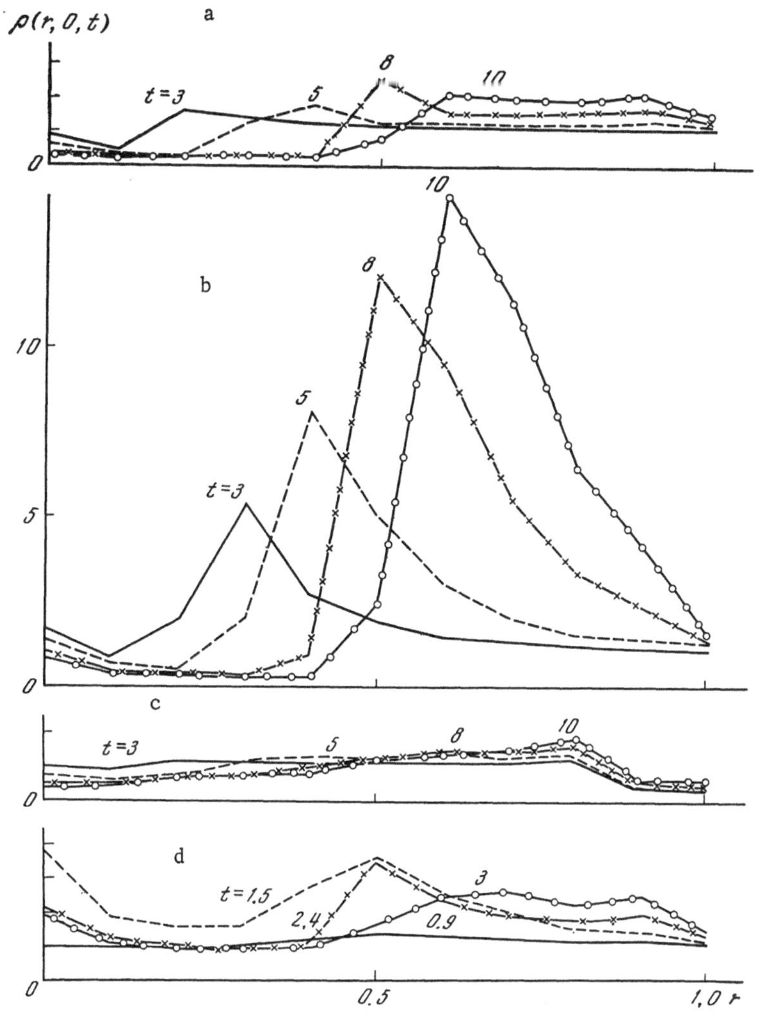

Fig. 4. Density profile ρ (r, 0, t) along the horizontal axis (φ = 0) at various times (the curves are labeled with the time. a) ν = 0, κ = 0.03, r_s = 0.1; b) ν = 0, κ = 0.03, r_s = 0.01; c) ν = 0, κ = 0.03, r_s = 0.3; d) ν = 0, r_s = 0.1, κ = 0.1. The ordinate scale is the same everywhere.

We used these parameters to calculate the vector potential A, the velocity components v_r and v_φ, the density ρ, and the current density j as functions of the coordinates and the time.

For the case of an ideal conductivity (ν = 0) the calculations were carried out at ten points along r and ten points along φ; the results are shown in Figs. 2-5. Figure 2 shows patterns of the magnetic lines of force at successive times. The lines of force tend to "pave" the horizontal axis, i.e., they tend to be compressed against the horizontal axis, forming a current sheet there. To estimate the length of this sheet, i.e., the distance to the "break" at the limiting line of force A = 0, near which the rapid deviation from the horizontal axis begins, we can use

$$x_c \simeq \sqrt{\varkappa t}. \tag{22}$$

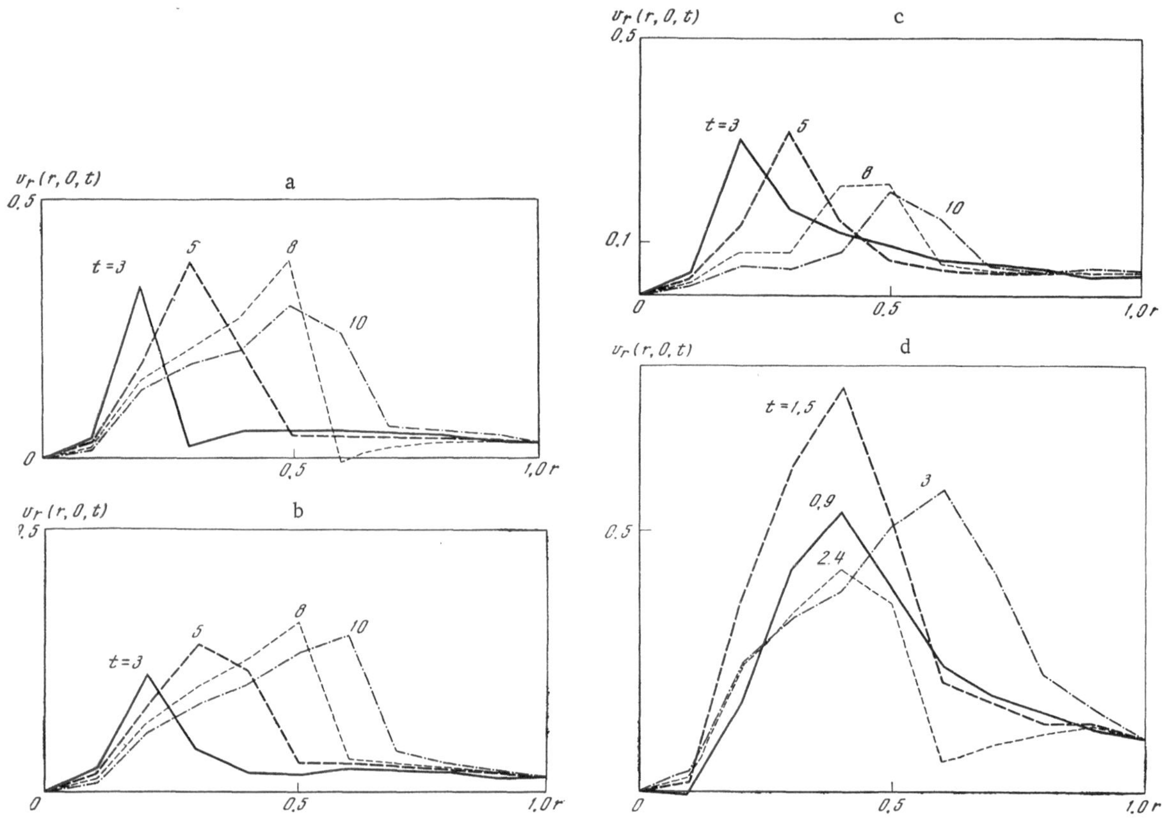

Fig. 5. Radial component of the velocity $v_r(r, 0, t)$ along the horizontal axis. a) $\nu = 0$, $\varkappa = 0.03$, $r_s = 0.1$; b) $\nu = 0$, $\varkappa = 0.03$, $r_s = 0.01$; c) $\nu = 0$, $\varkappa = 0.03$, $r_s = 0.3$; d) $\nu = 0$, $r_s = 0.1$, $\varkappa = 0.1$.

The sheet turns out to be very narrow. If we define the sheet thickness as the distance over which the magnetic field decays rapidly to zero, this thickness turns out to be (at any rate) smaller than a single spacing in the calculation scheme. This result can be seen easily from Fig. 3 which shows the horizontal component of the magnetic field, $H_\varphi(r, \pi/2, t)$, at the vertical axis, which passes through the null point.

The results for the case $\varkappa = 0.03$ and $r_s = 0.1$ are shown in Figs. 2 and 3; the patterns of lines of force obtained for other values of $\varkappa$ and r_s over the ranges in (20) and (21) are completely analogous, except that at large values of $\varkappa$ and r_s the sheet is initially less well defined: The dense plasma does not manage to flow away from the sheet region.

From Fig. 4, which shows the profile of the plasma density along the horizontal axis, including the sheet region, we see that the plasma density is lower than its original value, especially at small values of r_s. At the edge of the sheet, on the other hand, there is a high-density region, which moves away from the original null point as the sheet expands. We thus see that near the current sheet the plasma density decreases, in accordance with Syrovatskii's conclusion [4] (the equation given in [4] for estimating the extent of the plasma density decrease, which was actually justified only for the linear approximation, does not give the correct results for the nonlinear stage). We also note that the densities in Fig. 4 refer not to the sheet itself, which is very thin here, but to its surroundings. More precisely, these density values are averages over a region, one calculation spacing in size, which includes the current sheet and its neighborhood.

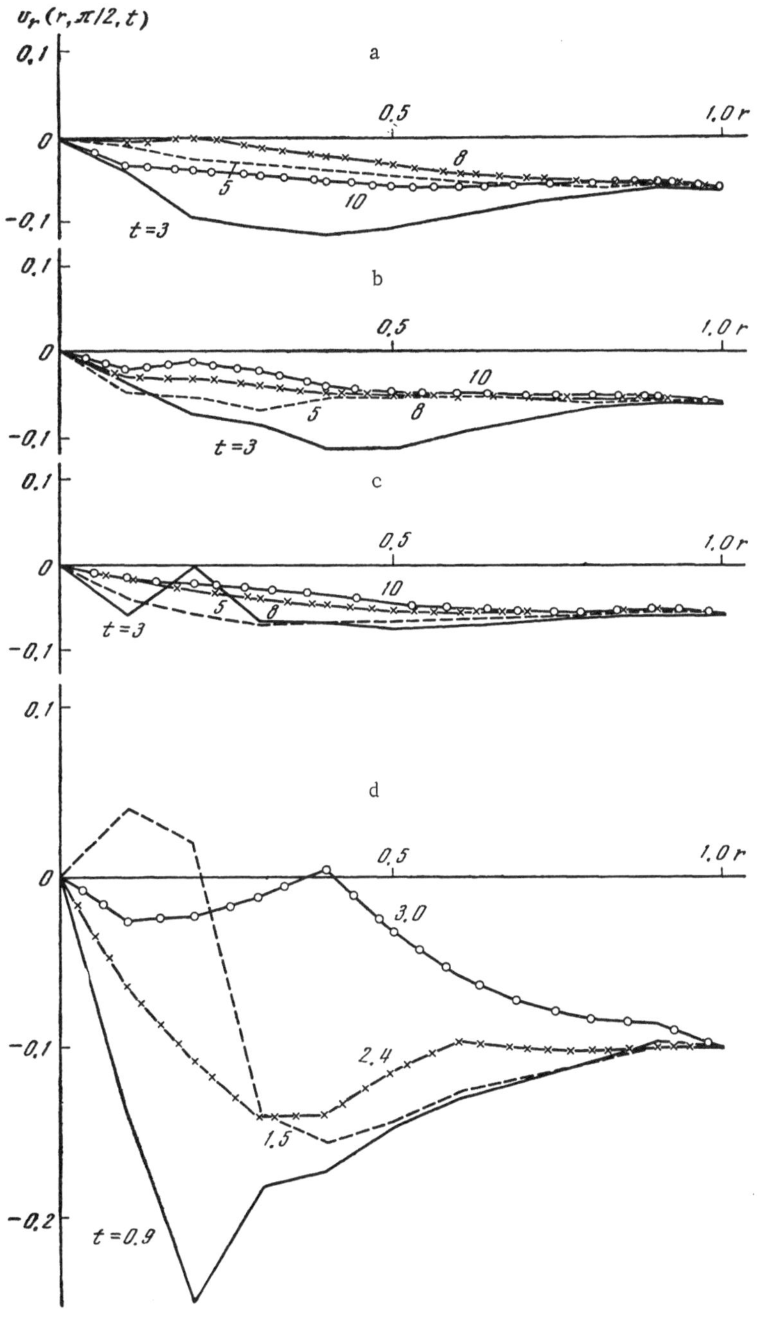

Fig. 6. Profile of the vertical velocity component v_r (r, $\pi/2$, t) along the vertical axis. a) $\nu = 0$, $\varkappa = 0.03$, $r_s = 0.1$; b) $\nu = 0$, $\varkappa = 0.03$, $r_s = 0.01$; c) $\nu = 0$, $\varkappa = 0.03$, $r_s = 0.3$; d) $\nu = 0$, $r_s = 0.1$, $\varkappa = 0.1$.

Figure 5 shows the velocity along the sheet, v_r (r, 0, t), at various times. We see that the plasma flows away from the sheet at a high velocity, reaching ~(0.3-0.7)V_A. The ejection of plasma at such a velocity is due primarily to electromagnetic forces in the sheet, rather than to the pressure gradient, since this gradient is small at the sheet and even positive near the edges of the sheet (i.e., it should retard the plasma motion there).

Figure 6 shows the profile of the vertical velocity component v_r (r, $\pi/2$, t) along the ray $\varphi = \pi/2$. We see that the plasma is moving toward the sheet slowly and that its velocity can even change sign, corresponding to the appearance of a wave reflected from the current sheet.

In the idealized problem treated here ($\nu = 0$) we were not able to follow the structural evolution of the sheet itself or the consequences of magnetic-energy dissipation in the sheet. We therefore carried out calculations for a second class of problems, in which we assumed $\nu \neq 0$, specifically, $\nu = \nu_0/\sqrt{\rho}$. We chose this latter relation because in a collisionless plasma the coefficient ν [see (6)] is governed by the turbulent plasma conductivity σ, which is frequently proportional to the square root of the plasma density. We chose the value $\nu_0 = 0.001$ in order to avoid severe "blurring" of the current sheet by the viscosity but to also avoid having to use a smaller time step in the local calculation scheme (because of too small a value of ν_0). The calculations were carried out for a grid consisting of 80 points along r and 40 points along φ.

Figure 7 shows patterns of the lines of force A(r, t, φ) = const for the values $\nu = 0.001$, $\varkappa = 0.1$, and $r_s = 0.1$ at various times. We see that, as in the case $\nu = 0$ (Fig. 2), a current

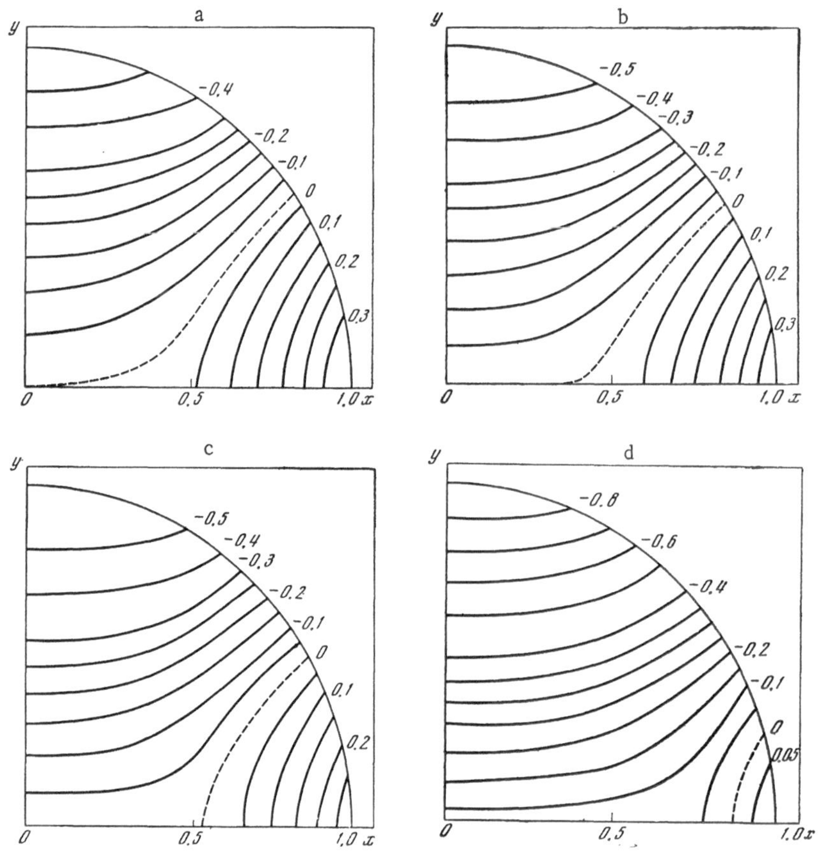

Fig. 7. Lines of force A(r, φ, t) = const for $\nu_0 = 0.001$, $\varkappa = 0.1$, and $r_s = 0.1$ at several times: a) t = 1.2; b) 1.5; c) 2.0; d) 4.0.

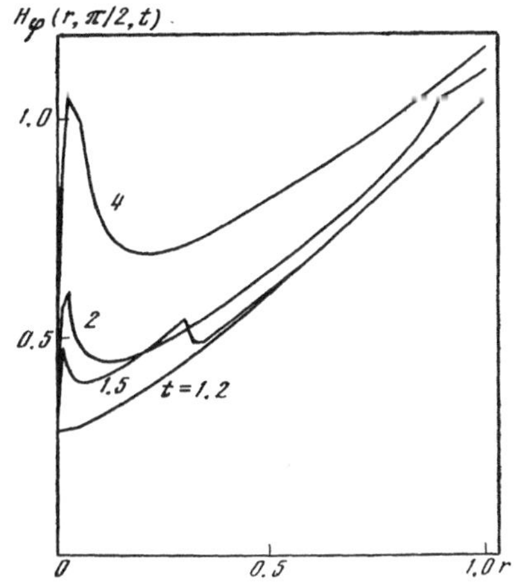

Fig. 8. Horizontal component of the magnetic field $H_\varphi(r, \pi/2, t)$ for $\nu_0 = 0.001$, $\varkappa = 0.1$, and $r_s = 0.1$.

sheet appears and becomes thicker, but, in contrast with the case of an infinite conductivity, there is a "reconnection" of the magnetic lines of force, as can be seen particularly clearly for the lines of force corresponding to A = 0 and −0.05.

These results verify Syrovatskii's conclusion [12] that a neutral current sheet forms as plasma moves along a null line. On the other hand, these results show no evidence for the appearance near this null line of the system of slow-mode shock waves proposed by Petschek [3]. Accordingly, in the contest between a neutral sheet and a system of shock waves the numerical calculations support the neutral sheet.

Figure 8 shows the time evolution of the profile of the magnetic field over the sheet thickness; we see that the field becomes much stronger near the sheet and varies rapidly within the sheet.

Figures 9 and 10 show the profile of the plasma density within the sheet ($\varphi = 0$) and near it, on the ray $\varphi = \pi/20$. Within the sheet the density has increased, especially at the center and ends of the sheet. Just outside the sheet, however, the plasma density has decreased

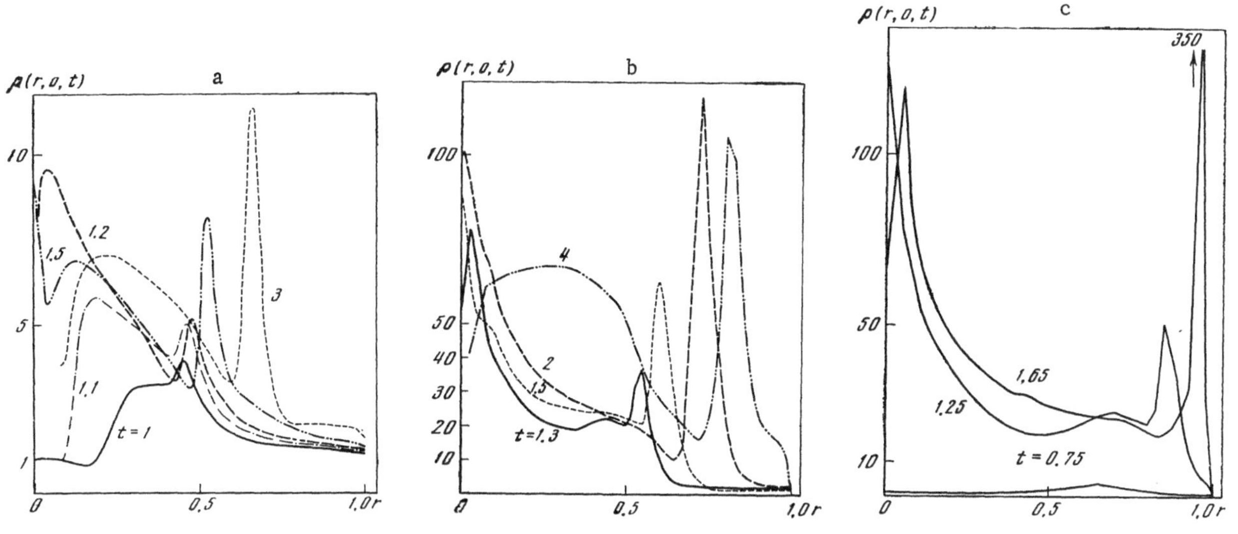

Fig. 9. Profile of the plasma density $\rho(r, 0, t)$ along the horizontal axis. a) $\nu_0 = 0.001$, $\varkappa = 0.1$, $r_s = 0.1$; b) $\nu_0 = 0.001$, $\varkappa = 0.1$, $r_s = 0.01$; c) $\nu_0 = 0.001$, $\varkappa = 0.2$, $r_s = 0.01$.

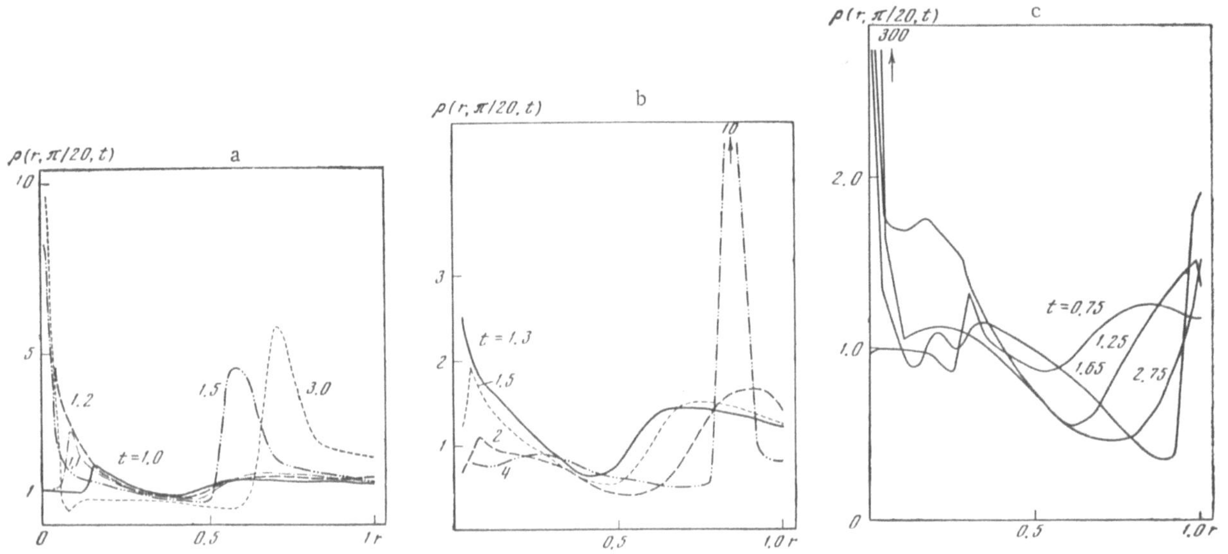

Fig. 10. Profile of the plasma density $\rho(r, \pi/20, t)$ near the sheet on the ray $\varphi = \pi/20$ at various times. a) $\nu_0 = 0.001$, $\varkappa = 0.1$, $r_s = 0.1$; b) $\nu_0 = 0.001$, $\varkappa = 0.1$, $r_s = 0.01$; c) $\nu_0 = 0.001$, $\varkappa = 0.2$, $r_s = 0.01$.

(Fig. 10a, b, c). Although this decrease is not extreme, we see that it is becoming more pronounced as time elapses, so that a significant density decrease can be expected after a long time. This conclusion supports the results of [13], where the calculations could be carried out for much longer times because of certain simplifications possible for the case of an established current sheet. In this regard the results of the present study can be thought of as a justification for the assumption in [13] of the appearance of a current sheet at a null line.

Figure 11 shows that the plasma flows away from the sheet at a velocity amounting to a few tenths of the Alfvén velocity. This result may be of interest in connection with the theory of solar flares, in which plasma ejections at velocities approaching the Alfvén velocity are frequently observed [14].

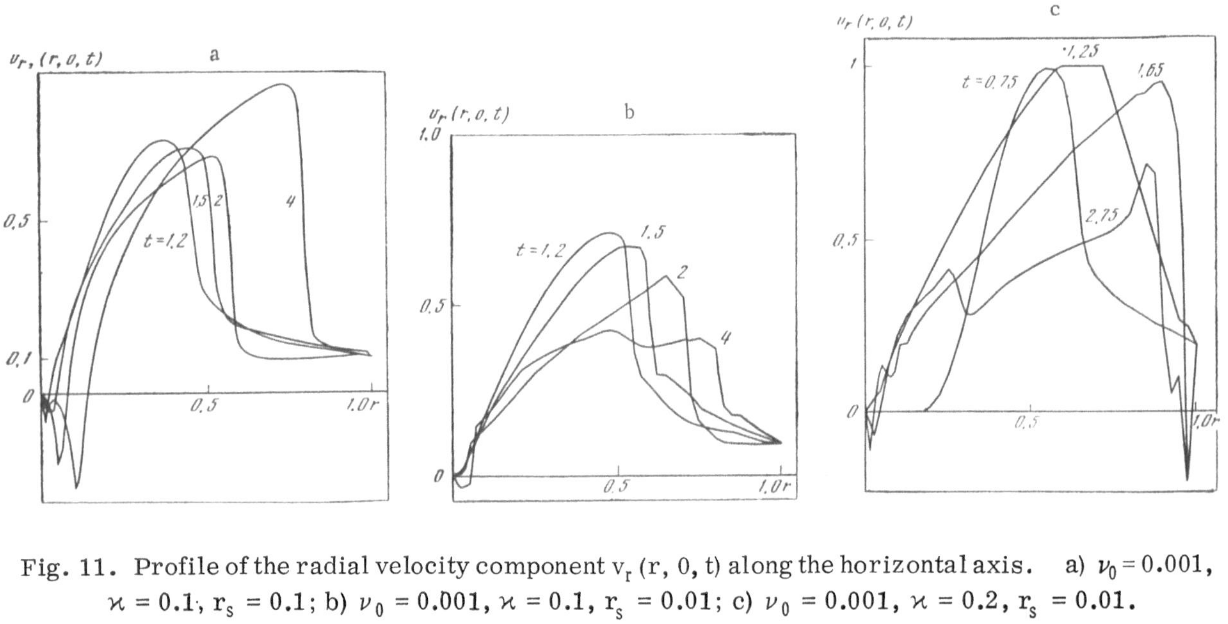

Fig. 11. Profile of the radial velocity component $v_r(r, 0, t)$ along the horizontal axis. a) $\nu_0 = 0.001$, $\varkappa = 0.1$, $r_s = 0.1$; b) $\nu_0 = 0.001$, $\varkappa = 0.1$, $r_s = 0.01$; c) $\nu_0 = 0.001$, $\varkappa = 0.2$, $r_s = 0.01$.

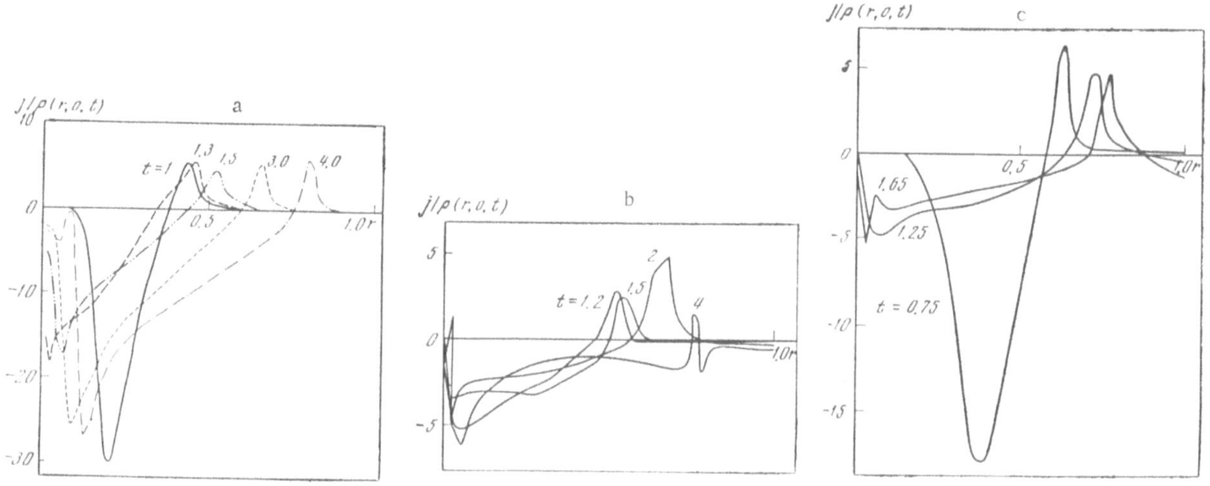

Fig. 12. Profile of the specific current density j/ρ (r, 0, t) along the horizontal axis. a) $\nu_0 = 0.001$, $\varkappa_0 = 0.1$, $r_s = 0.1$; b) $\nu_0 = 0.001$, $\varkappa = 0.1$, $r_s = 0.01$; c) $\nu_0 = 0.001$, $\varkappa = 0.2$, $r_s = 0.01$.

Finally, Fig. 12 shows the profile of the specific current density $j(r, 0, t)/\rho(r, 0, t)$ along the current sheet. We clearly see a region of a return current at the edge of the sheet. The possible appearance of such regions in the case of sheets developing sufficiently rapidly was demonstrated in [12].

We can summarize the basic results of this study as follows:

1. A change in the field sources causes a converging cylindrical perturbation wave near a null line.

2. As it approaches the null line this wave becomes nonlinear and leads to the appearance of a thin current sheet, which becomes progressively wider.

3. There is no evidence for the system of slow shock waves postulated by Petschek near the null line.

4. Within the current sheet and at its edges the plasma density becomes higher than its original value.

5. The plasma density falls below its original value in the region adjacent to both surfaces of the sheet.

6. The plasma flows away from the sheet during the sheet formation at a velocity reaching several tenths of the Alfvén velocity.

7. Regions of a return current (a current flowing in the opposite direction) exist at the edges of the sheet.

The authors thank V. S. Imshennik and O. V. Lokutsievskii for useful discussions.

Literature Cited

1. J. W. Dungey, Cosmic Electrodynamics, Cambridge Univ. Press (1958).
2. P. A. Sweet, Annual Review of Astronomy and Astrophysics, Vol. 7 (1969), p. 149.
3. H. E. Petschek, Proceedings of the AAS-NASA Symposium on the Physics of Solar Flares, Washington, 1964, p. 426.
4. S. I. Syrovatskii, Astron. Zh. 43:340 (1966).

5. S. I. Syrovatskii, Zh. Éksp. Teor. Fiz., 50:1133 (1966).
6. S. I. Syrovatskii, in: Solar Flares and Space Research (ed. Z. Svestka and C. de Jager), North-Holland, Amsterdam (1969), p. 346.
7. S. I. Syrovatskii, in: Solar-Terrestrial Physics, 1970 (E. R. Dyer, ed.), Reidel, Dordrecht (1972), p. 119.
8. V. S. Imshennik and S. I. Syrovatskii, Zh. Éksp. Teor. Fiz., 52:990 (1967).
9. S. Chapman and P. C. Kendall, Proc. Roy. Soc., A271:485 (1963).
10. S. I. Syrovatskii, Zh. Éksp. Teor. Fiz., 54:1422 (1968).
11. J. C. Stevenson, J. Plasma Phys., 7:293 (1972).
12. S. I. Syrovatskii, Zh. Éksp. Teor. Fiz., 60:1727 (1971).
13. B. V. Somov and S. I. Syrovatskii, Tr. FIAN, 74:14 (1974).
14. C. de Jager, "Structure and dynamics of the solar atmosphere," Handbuch der Physik, Vol. LII, Astrophysics III: The Solar System (S. Flügge, ed.), Springer-Verlag, Berlin (1959), p. 80.

KINETICS OF A NEUTRAL CURRENT SHEET

S. V. Bulanov and S. I. Syrovatskii

In an analysis of the propagation of various types of waves in a thin neutral current sheet, dispersion relations are obtained in the kinetic approximation. Streaming instabilities and the tearing-mode instability in neutral sheets are analyzed. In the nonrelativistic approximation, a current sheet in an unsteady external electric field directed along the current is unstable against tearing; in the ultrarelativistic approximation the current sheet is stable against tearing if the external field increases sufficiently rapidly ($E_0 \sim t^m$, $m > 1$). Three simple models for the acceleration of charged particles in neutral fields are worked out: acceleration of particles in the sheet and the surrounding plasma during a rapid rupture of the current sheet, and acceleration of particles when a strong electromagnetic pulse is incident on the sheet. A simple model describing the rupture of the current sheet is analyzed. It is shown that the electric field which arises at the rupture has an amplitude $E \sim (v/c)H_0$, where H_0 is the external magnetic field and v is the characteristic plasma velocity in the sheet.

Introduction

The interest in neutral current sheets in plasma in space and in the laboratory results from the possibility that these sheets may furnish an explanation for the rapid changes in magnetic field configurations in space and the acceleration of charged particles to high energies. A study of neutral current sheets necessarily includes a study of the conditions for the appearance and existence of current sheets and a study of their decay, during which magnetic energy is liberated and particles are accelerated.

The appearance of current sheets was studied in [1-10]. It has been shown [1-4] that in the MHD approximation, in the approximation of a strong field, $H^2/8\pi \gg \rho v^2/2, n\varkappa T$ [1, 3, 4], current sheets form near points where the magnetic field vanishes but where the electric field does not vanish.

Calculations carried out [3, 5] for plasma flow near a current sheet yield a very important result: During the growth of a current sheet the plasma density near the sheet decays without bound and can reach values satisfying the condition

$$H^2/4\pi nmc^2 \gg 1. \tag{1}$$

In this case a neutral sheet is a plasma sheet confined by a magnetic field, which is in turn produced by the current flowing in the sheet [11]. Such a system is unstable. Various types of instabilities can arise in it; for example, there is the ion-acoustic instability, which leads to heating and to the onset of turbulence in the sheet [10, 12, 34, 35], and there is the electromagnetic tearing-mode instability [13-23].

In the present paper we will be concerned with the stability and with simple models for the decay dynamics of the current sheet and the plasma near it.

1. Equilibrium State

By "neutral current sheet" here we mean a set of charged particles which produce a plane current (near the y = 0 plane) which separates regions in which the magnetic fields are equal in magnitude but opposite in sign (Fig. 1). At y = 0, i.e., at the center of the sheet, the field vanishes. The total current in the sheet, I, is unambiguously related to the magnetic field $|H| = H_0 = 2\pi I/c$ far from the sheet, at $y = \pm\infty$. The plasma carrying the current is confined by the field of this current. A self-consistent problem of this type for a collisionless plasma was solved by Harris [11]. The distribution function for particles of species j = e, i in the sheet is

$$f_{0j} = \frac{n_j \exp\left[-(v_x^2 + v_y^2)/2\bar{v}_j^2 - (v_z - V_j)^2/2\bar{v}_j^2\right]}{(2\pi\bar{v}_j^2)^{3/2} \cosh^2(y/L)}. \tag{1.1}$$

The particle distribution in the sheet depends on the coordinate y only:

$$n_j(y) = n_j / \cosh^2(y/L). \tag{1.2}$$

The sheet thickness L is $(T_j = m_j\bar{v}_j^2/2\varkappa)$

$$L = \left[\frac{\varkappa T_i c^2}{2\pi n e^2 V_i^2 (1 + T_e/T_i)}\right]^{1/2}, \tag{1.3}$$

where V_j is the directed velocity of the particles in the sheet.

The y dependence of the magnetic field is

$$H(y) = \sqrt{8\pi n\varkappa (T_e + T_i)} \tanh(y/L). \tag{1.4}$$

Equations (1.1)-(1.4) were derived for a current sheet in which there is no charge separation and in which the condition

$$\frac{V_i}{T_i} = -\frac{V_e}{T_e} \tag{1.5}$$

holds. Then Eqs. (1.1)-(1.4) are valid for a sheet of arbitrary thickness, including a thickness which is smaller than the particle's gyroradius.

However, the question arises of whether charge separation in the sheet can legitimately be neglected if $L \lessapprox r_{H_i}$, i.e., if the thickness of the sheet is on the order of or less than the ion gyroradius. In such a neutral sheet the electrons are still magnetized but the ions are not,

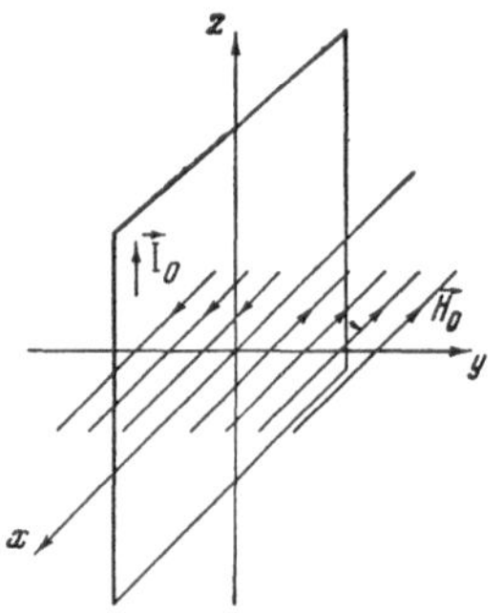

Fig. 1. Current sheet and coordinate system.

and a plasma polarization can apparently arise. A question of considerable interest in this connection (and a question which has not yet been studied) is that of whether condition (1.5) can be maintained in the presence of real collisions or "effective" collisions, due to plasma turbulence in the sheet.

2. Stability of a Neutral Current Sheet

There have been several papers on the stability of neutral current sheets [12-23]; the most interesting instability for the problem of particle acceleration is the anisotropic electromagnetic tearing-mode instability [13-21]. As a result of the onset of this instability the current sheet breaks up into filaments which are homogeneous along the direction of the current in the sheet, and the magnetic lines of force reconnect through the sheet. At the point of the rupture the plasma density decreases and electric fields arise. Such regions are interesting as a possible site of particle acceleration in a plasma.

In addition to this instability there is considerable interest in the various streaming instabilities (the ion-acoustic instability, etc.), which can cause turbulence in the sheet and produce a resistance limiting the current in the plasma [10, 12, 34, 35].

The tearing-mode instability has been analyzed in various approximations, and it has been shown that this instability occurs both in the case of a collisionless plasma and in the MHD approximation. It occurs when the finite plasma conductivity due to collisions is taken into account or when the inertial term in the generalized Ohm's law is taken into account in the case of a collisionless plasma. We note in this connection that sometimes (see, e.g., [21a]) the dissipation due to collisions or plasma turbulence has been treated as a necessary condition for the onset of the tearing mode. Actually, the tearing mode occurs in either a collisional or collisionless plasma and can itself be thought of as a mechanism tending to dissipate the energy of the current sheet.

In all cases the condition for the onset of the tearing-mode instability is [13-16]

$$kL < 1; \tag{2.1}$$

i.e., the perturbation wavelength $\lambda = 2\pi/k$ must be larger than $2\pi L$, where L is the sheet thickness. This instability is aperiodic. In the MHD approximation [15] its growth rate is

$$\gamma \simeq \tau_R^{-3/5}\tau_A^{-2/5}(kL)^{-2/5}, \tag{2.2}$$

where $\tau_R = 4\pi\sigma L^2/c^2$ is the scale time for diffusion of the magnetic field across the sheet due to the finite conductivity σ, $\tau_A = L/v_A$, and $v_A = H/(4\pi n M)^{1/2}$ is the Alfvén velocity.

In a collisionless plasma, if the sheet is thick, i.e., if the thickness L is larger than the particle's gyroradius $r_{Hj} = m_j \bar{v}_j c/eH$ in the sheet, where j = e, i, the growth rate is [17-20]

$$\gamma \simeq \frac{k\bar{v}_e}{\sqrt{\pi}}\left(\frac{r_{H_e}}{L}\right)^{3/2}\left(1+\frac{T_i}{T_e}\right). \tag{2.3}$$

The nonlinear stage of the growth of the tearing-mode instability was analyzed in the quasilinear approximation under the condition $L \gg r_{Hj}$ in [22, 23]; the results reveal a strong nonlinear stabilization of this instability and that the energy of the turbulent fluctuations increases with decreasing sheet thickness. In the case of a thin sheet the instability is presumably not stabilized and leads to important nonlinear effects: magnetic-field annihilation and effective acceleration of charged particles. Accordingly, the instability of a thin ($L < r_{Hj}$) neutral current sheet is of particular interest in connection with the problem of charged-particle acceleration.

The stability of a thin neutral current sheet against various types of perturbations was analyzed in [13, 24]. Below we attempt to analyze the stability of a thin sheet in more detail.

We assume that the current sheet is initially ($t = 0$) in the $y = 0$ plane. A current $I = e(n_{0i}V_i - n_{0e}V_e)$ flows along the z axis in the sheet. The magnetic field is $\mathbf{H} = H_0$ (sign y, 0, 0), where $H_0 = 2\pi I/c$, and n_{0i} and n_{0e} are the numbers of ions and electrons, respectively, per unit surface area (per square centimeter) of the sheet. The unperturbed distribution function for particles of species j, which can be found from Eq. (1.1) by taking the limit as the current sheet vanishes at constant $n_0 = \int_{-\infty}^{+\infty} n(y)\,dy$, is

$$f_j(y, \mathbf{v}) = \delta(y) f_{0j}(\mathbf{v}) = \frac{n_{0j}\delta(y)}{2\pi\bar{v}_{xj}\bar{v}_{zj}} \exp\left[-v_x^2/2\bar{v}_{xj}^2 - (v_z - V_j)^2/2\bar{v}_{zj}^2\right]. \tag{2.4}$$

In general we assume $\bar{v}_{xj} \neq \bar{v}_{zj}$. We further assume $n_{0e} = n_{0i} = n_0$. The sheet is in vacuum or in a plasma with a density n, with a dielectric constant

$$\varepsilon_0(\omega) = 1 - \frac{\omega_p^2}{(\omega + i\nu)^2}. \tag{2.5}$$

Here $\omega_p = (4\pi ne^2/m)^{1/2}$ and ν is the effective collision frequency. We are assuming symmetry with respect to y.

We write the coordinate and time dependence of perturbed properties, denoted by a subscript "1," in the plane of the sheet as $\sim\exp(-i\omega t + ik_x x + ik_z z)$. We also introduce the potentials $\mathbf{A}_1$ and φ_1:

$$\mathbf{E}_1 = -\nabla\varphi_1 - \frac{1}{c}\frac{\partial \mathbf{A}_1}{\partial t}, \tag{2.6}$$

$$\mathbf{H}_1 = \operatorname{rot} \mathbf{A}_1. \tag{2.7}$$

The perturbed distribution of particles of species j in the sheet is

$$f_{1j}(\omega, \mathbf{k}, y, \mathbf{v}) = \left[\frac{e_j}{m_j c}\left(\mathbf{A}_1 \cdot \frac{\partial f_{0j}}{\partial \mathbf{v}}\right) - \frac{e_j}{m_j} \frac{\varphi_1 + \frac{1}{c}(\mathbf{A}_1 \cdot \mathbf{v})}{\omega - (\mathbf{k} \cdot \mathbf{v})}\left(\mathbf{k} \cdot \frac{\partial f_{0j}}{\partial \mathbf{v}}\right)\right]\delta(y). \tag{2.8}$$

Calculating the perturbed values of the current density $\mathbf{j}_1$ and the charge ρ_1 in the standard manner we find the following equations for the potentials:

$$\frac{d^2\mathbf{A}_1}{dy^2} - \left(k_x^2 + k_z^2 - \varepsilon_0(\omega)\frac{\omega^2}{c^2}\right)\mathbf{A}_1 =$$
$$= -\sum_j \frac{4\pi e_j^2}{m_j c^2} \int d^2v\,\mathbf{v}\left[\left(\mathbf{A}_1 \cdot \frac{\partial f_{0j}}{\partial \mathbf{v}}\right) - \frac{c\varphi_1 + (\mathbf{A}_1 \cdot \mathbf{v})}{\omega - (\mathbf{k} \cdot \mathbf{v})}\left(\mathbf{k} \cdot \frac{\partial f_{0j}}{\partial \mathbf{v}}\right)\right]\delta(y), \tag{2.9}$$

$$\frac{d^2\varphi_1}{dy^2} - \left(k_x^2 + k_z^2 - \varepsilon_0(\omega)\frac{\omega^2}{c^2}\right)\varphi_1 =$$
$$= -\sum_j \frac{4\pi e_j^2}{m_j c} \int d^2v\left[\left(\mathbf{A}_1 \cdot \frac{\partial f_{0j}}{\partial \mathbf{v}}\right) - \frac{c\varphi_1 + (\mathbf{A}_1 \cdot \mathbf{v})}{\omega - (\mathbf{k} \cdot \mathbf{v})}\left(\mathbf{k} \cdot \frac{\partial f_{0j}}{\partial \mathbf{v}}\right)\right]\delta(y). \tag{2.10}$$

This system of equations has the following solution:

$$A_1(\omega, \mathbf{k}, y) = \frac{2\pi \exp\left(-|y|\sqrt{k^2 - \varepsilon_0(\omega)\omega^2/c^2}\right)}{c^2\sqrt{k^2 - \varepsilon_0(\omega)\omega^2/c^2}} \sum_j \frac{e_j^2}{m_j} \int d^2v\mathbf{v}\left[\left(\mathbf{A}_1(\omega, \mathbf{k}, y=0)\frac{\partial f_{0j}}{\partial \mathbf{v}}\right) - \right.$$
$$\left. - \frac{c\varphi_1(\omega, \mathbf{k}, y=0) + (\mathbf{A}_1(\omega, \mathbf{k}, y=0)\,\mathbf{v})}{\omega - (\mathbf{k}\cdot\mathbf{v})}\left(\mathbf{k}\cdot\frac{\partial f_{0j}}{\partial \mathbf{v}}\right)\right], \tag{2.11}$$

$$\varphi_1(\omega, \mathbf{k}, y) = \frac{2\pi \exp\left(-|y|\sqrt{k^2 - \varepsilon_0(\omega)\omega^2/c^2}\right)}{c\sqrt{k^2 - \varepsilon_0(\omega)\omega^2/c^2}} \sum_j \frac{e_j^2}{m_j} \int d^2v\left[\left(\mathbf{A}_1(\omega, \mathbf{k}, y=0)\frac{\partial f_{0j}}{\partial \mathbf{v}}\right) - \right.$$
$$\left. - \frac{c\varphi_1(\omega, \mathbf{k}, y=0) + (\mathbf{A}_1(\omega, \mathbf{k}, y=0)\,\mathbf{v})}{\omega - (\mathbf{k}\cdot\mathbf{v})}\left(\mathbf{k}\frac{\partial f_{0j}}{\partial \mathbf{v}}\right)\right]. \tag{2.12}$$

Here $k^2 = k_x^2 + k_z^2$.

Setting y = 0 in Eqs. (2.11) and (2.12) we find a system of linear algebraic equations for $A_{1x}(\omega, \mathbf{k}, y = 0)$, $A_{1z}(\omega, \mathbf{k}, y = 0)$, and $\varphi_1(\omega, \mathbf{k}, y = 0)$. Finally, equating the determinant of this system to zero, we find the dispersion relation for waves in the neutral sheet.

We turn now to several particular cases.

1. For plasma waves in the sheet we assume $\mathbf{A}_1 = 0$, $\mathbf{k} = (k_x, 0)$. For a sheet in vacuum we would have n = 0. The dispersion relation for these waves is

$$\varepsilon(\omega, k) = 1 + \sum_j \frac{\omega_{0j}c}{k\bar{v}_j^2}\left[1 + i\sqrt{\pi}\frac{\omega}{k\bar{v}_j}W\left(\frac{\omega}{k\bar{v}_j}\right)\right]. \tag{2.13}$$

where

$$\omega_{0j} = \frac{2\pi n_0 e_j^2}{m_j c} \tag{2.14}$$

is the scale frequency in the thin neutral sheet, W(x) can be expressed in terms of the probability integral of complex argument [25, 26], and we are assuming electrostatic waves, $\omega \ll kc$.

We see that dispersion relation (2.13) for waves in a thin plasma sheet is similar to that for plasma waves in a homogeneous plasma, but the coefficient of the expression in brackets in (2.13) is different. For a homogenous plasma it would be $(k^2L_{Dj}^2)^{-1}$ [27, 28], while for a thin neutral sheet it is $\omega_{0j}c/k\bar{v}_j^2$.

The analogs of plasma waves in a "cold" plasma, $\omega \gg k\bar{v}_e$, are waves for which the real part of the frequency is (see also [29])

$$\omega_k = \sqrt{\omega_{0e}ck}. \tag{2.15}$$

We see that, in constrast with a homogeneous plasma, in which ω does not depend on k for such waves, the phase velocity for these waves in the present case is $v_\varphi = (\omega_{0e}c/k)^{1/2}$. The growth rate for the waves is

$$\gamma_k = -\frac{\omega_{0e}^2c^2}{2\sqrt{\pi}k\bar{v}_e^3}\exp\left(-\omega_{0e}c/k\bar{v}_e^2\right), \tag{2.16}$$

as can be seen easily by making use of the standard results for a homogeneous plasma.

2. The analogs of ion-acoustic waves, with $T_i \ll T_e$, are the waves described by the dispersion relation

$$\varepsilon(\omega, k) = 1 + \frac{\omega_{0e}c}{k\bar{v}_e^2}\left[1 + \frac{i\sqrt{\pi}\omega}{k\bar{v}_e}\right] + \frac{\omega_{0i}c}{k\bar{v}_i^2}\left[1 + i\sqrt{\pi}\frac{\omega}{k\bar{v}_i}W\left(\frac{\omega}{k\bar{v}_i}\right)\right] = 0. \tag{2.17}$$

If $\omega \gg k\bar{v}_i$, the square frequency of these waves is

$$\omega_k^2 = \frac{\omega_{0i}ck}{(1+\omega_{0e}c/k\bar{v}_e^2)}\left[1+3\frac{k\bar{v}_i^2}{\omega_{0i}c}\left(1+\frac{\omega_{0e}c}{k\bar{v}_e^2}\right)\right]; \quad (2.18)$$

if $k < \omega_{0e}c/\bar{v}^2$, it is

$$\omega_k^2 = k^2\frac{\varkappa T_e}{M}\left(1+3\frac{T_i}{T_e}\right). \quad (2.19)$$

In other words, the dispersion of these waves is the same as that in a homogeneous plasma [27].

3. By analogy with the ion plasma waves we have

$$\omega_k = \sqrt{\omega_{0i}ck}. \quad (2.20)$$

This equation is essentially the same as Eq. (2.15); however, such waves require a stringent condition on k:

$$k > \omega_{0e}c/\bar{v}_e^2 \simeq \frac{1}{L}\left(\frac{c}{V}\right)^2, \quad (2.21)$$

and in this case the sheet can no longer be assumed thin [see (1.3)].

4. In the case of streaming instabilities in a current sheet the electrons drift at a velocity $V_i + V_e$ with respect to the ions. Under these conditions we could find the instabilities which arise when there is a relative motion of electrons and ions [24, 27]. In Eqs. (2.11), (2.12) this case corresponds to waves with $A_1 = 0$, $k = (0, 0, k_z)$. We further assume that $\varepsilon_0(\omega) = 1$ and $\omega \ll kc$.

The dispersion relation for these waves is

$$\varepsilon(\omega, k) = 1 + \sum_j \frac{\omega_{0j}c}{k_z\bar{v}_{zj}^2}\left[1 + i\sqrt{\pi}\frac{(\omega - k_zV_j)}{k_z\bar{v}_{zj}}W\left(\frac{\omega - k_zV_j}{k_z\bar{v}_{zj}}\right)\right] = 0. \quad (2.22)$$

As above, exploiting the analogy between the dispersion relations in a homogeneous plasma and a current sheet, we find that in a "cold" plasma, $|\omega - kV_j| \gg k\bar{v}_j$, there is an analog of the Buneman instability. The condition for the onset of this instability in the coordinate system tied to the ions [i.e., the condition for a resonance between ion drift waves $\omega \simeq 0$, and waves of the type in (2.15)] is

$$|k| < \frac{\omega_{0e}c}{V^2}. \quad (2.23)$$

The sheet can be assumed thin if $V^2 \gg v_e c$. The growth rate near the threshold in (2.23) is

$$\gamma = \sqrt{\omega_{0e}ck}\,\frac{\sqrt{3}}{2^{4/3}}\left(\frac{m}{M}\right)^{1/3}\omega_{0e}\frac{c}{V}\frac{\sqrt{3}}{2^{1/3}}\left(\frac{m}{M}\right)^{1/3}. \quad (2.24)$$

The real part of the frequency is

$$\omega = -\frac{1}{2^{4/3}}\sqrt{\omega_{0e}ck}\left(\frac{m}{M}\right)^{1/3}. \quad (2.25)$$

5. The condition for the onset of the ion-acoustic instability, $T_e \gg T_i$, is (for $k < \omega_{0e}c/\bar{v}_e^2$)

$$V > \sqrt{\frac{\varkappa T_e}{M}} \quad (2.26)$$

and the corresponding growth rate is

$$\gamma \simeq \left(\frac{m}{M}\right)^{1/2} k \sqrt{\frac{\varkappa T_e}{M}}. \tag{2.27}$$

Let us examine the necessary conditions for the onset of turbulence in the current sheet. Using Eq. (1.3) for the thickness of the sheet as a function of V_j and v_j, and using condition (2.26), we find that if $T_e \gg T_i$ a sheet with a thickness

$$L \approx r_{H_i} \left(\frac{T_e}{T_i}\right)^{1/2} \tag{2.28}$$

becomes turbulent. If the ion and electron temperatures are comparable, a condition for the onset of the two-stream instability is [27]

$$V > \bar{v}_e; \tag{2.29}$$

i.e., if $T_e \approx T_i$ the sheet becomes turbulent if its thickness is

$$L \approx r_{H_e}. \tag{2.30}$$

6. Transverse electromagnetic waves in a homogeneous plasma, with a dispersion relation [30]

$$k^2c^2 - \omega^2 - i\sqrt{\pi} \sum_j \omega_{pj}^2 \frac{\omega}{k\bar{v}_j} W\left(\frac{\omega}{k\bar{v}_j}\right) = 0, \tag{2.31}$$

are analogous to waves in a thin plasma sheet described by the equation

$$1 - i\sqrt{\pi} \sum_j \frac{W\left(\frac{\omega}{k\bar{v}_j}\right) \omega_{0j}/k\bar{v}_j}{\lambda_{0j}\sqrt{k^2 - \varepsilon_0(\omega)\,\omega^2/c^2}} = 0. \tag{2.32}$$

Here $\lambda_{0j} = c/\omega_{0j} = m_j c^2/2\pi n_{0j} e_j^2$.

In a "cold" plasma, $\omega/kv \gg 1$, we have

$$k^2c^2 - \omega^2 + \frac{\omega^2\omega_p^2}{(\omega + i\nu)^2} - \omega_{0e}^2 = 0. \tag{2.33}$$

In contrast with a homogeneous plasma, in which [30, 31]

$$k^2c^2 - \omega^2 + \omega_p^2 = 0, \tag{2.34}$$

the waves described by Eq. (2.33) have an electric field component E_z and magnetic field components H_x and H_y, with $k \neq 0$. In the sheet (at $y = 0$) we have $H_x = 0$, and the properties E_z, H_y, k_x form a triad of mutually perpendicular vectors.

7. The electromagnetic tearing-mode instability of a neutral current sheet arises in the plasma if there is an anisotropic velocity distribution. Instabilities of this type can be more important than electrostatic instabilities if the ratio of the plasma pressure P_0 to the pressure of the static magnetic field, $H^2/8\pi$, is finite: $\beta \equiv 8\pi P_0/H^2 \geq 1$ [28]. This condition is satisfied in the central part of a neutral current sheet. The tearing-mode instability which arises in this case, as menioned above, is of particular interest in connection with the problem of particle acceleration at current sheets.

The general dispersion relations obtained from (2.11) and (2.12) describe the tearing-mode instability for the case of perturbations homogeneous in the current direction $\varphi_1 \neq 0$, $A_{1x} = 0$, and $A_{1z} \neq 0$.

For these waves we find the dispersion relation

$$\left(1+\sum_j \frac{1-(V_j^2+\bar{v}_{zj}^2)/\bar{v}_{xj}^2\,[1+i\sqrt{\pi}(\omega/k\bar{v}_{xj})W(\omega/k\bar{v}_{xj})]}{\lambda_{0j}\sqrt{k^2-\varepsilon_0(\omega)\,\omega^2/c^2}}\right)\times$$
$$\times\left(1+\sum_j \frac{c^2}{\lambda_{0j}\sqrt{k^2-\varepsilon_0(\omega)\omega^2/c^2\bar{v}_{xj}^2}}[1+i\sqrt{\pi}(\omega/k\bar{v}_{xj})W(\omega/k\bar{v}_{xj})]\right)+$$
$$+\left(\sum_j \frac{V_j c}{\lambda_{0j}\sqrt{k^2-\varepsilon_0(\omega)\,\omega^2/c^2\bar{v}_{xj}^2}}[1+i\sqrt{\pi}(\omega/k\bar{v}_{xj})W(\omega/k\bar{v}_{xj})]\right)^2=0. \qquad (2.35)$$

In the hydrodynamic approximation – for long wavelengths ($\omega \gg k\bar{v}_j$, $\omega \ll kc$) – this equation becomes

$$\left(1+\sum_j \frac{1+k^2V_j^2/\omega^2}{\lambda_{0j}k}\right)\left(1-\sum_j \frac{\omega_{0j}kc}{\omega^2}\right)+\left(\sum_j \frac{\omega_{0j}kV_j}{\omega^2}\right)^2=0. \qquad (2.36)$$

We assume an unperturbed ion velocity $V_i = 0$ and a directed electron velocity V_e. This equation was analyzed in [13].

The growth rate in this case is

$$\gamma=\frac{kV_e}{\sqrt{\lambda_{0e}k+1}}\left(\frac{m}{M}\right)^{1/2}. \qquad (2.37)$$

We see that at long perturbation wavelengths the growth rate decreases in inverse proportion to the perturbation wavelength. In analyzing the stability of a thin neutral sheet it is important to take the ion motion into account [13]. If we had neglected the ion motion in Eqs. (2.9) and (2.10) ($\varphi_1 = 0$), we would have found a growth rate for the "electron" tearing-mode instability larger by a factor of $(M/m)^{1/2}$ than that in (2.37).

8. Equation (2.37) does not hold for short wavelengths. In this case, for $\omega/k\bar{v}_j \ll 1$, we find from (2.35) the dispersion relation ($V_i = 0$)

$$\left(1-\frac{1}{kL}-i\sqrt{\pi}\frac{\omega}{k\bar{v}_e(kL)}\right)\left[1+\frac{2c^2}{(kL)\,V^2}+i\sqrt{\pi}\frac{\omega}{k\bar{v}_e}\frac{c^2}{(kL)\,V^2}\left(1+\frac{\bar{v}_e}{\bar{v}_i}\right)\right]+\frac{c^2}{(kL)^2\,V^2}\left(1+i\sqrt{\pi}\frac{\omega}{k\bar{v}_e}\right)^2=0. \qquad (2.38)$$

The condition for the stability of the sheet with respect to the tearing-mode instability is the same as before: $kL < 1$.

At $kL \approx 1$ the growth rate for this instability is

$$\gamma\simeq\frac{k\bar{v}_e}{\sqrt{\pi}}(1-kL). \qquad (2.39)$$

However, this expression is applicable only in a very narrow range of $k \sim L^{-1}$, since it was obtained in the approximation $\gamma/k\bar{v}_e$, $\gamma/k\bar{v}_i \ll 1$.

We turn now to the stability of a "thick" current sheet, $L \gg r_{Hj}$, against the tearing mode. An analogous problem was solved in [17-20]. For generality here we give its solution by the method used to analyze the stability of a thin current sheet. We take perturbations of the electrostatic potential φ_1 as well as of A_1 into account. In the approximation of a "thick" current sheet, $L \gg r_{Hj}$, the equations describing the perturbations in the case $|y| < L$ are

$$\left[\frac{d^2}{dy^2}-k^2+\frac{2}{L^2}\right]A_1=\delta(y)\sum_j 2\left[1-(V_j^2+\bar{v}_{zj}^2)/\bar{v}_{xj}^2\,[1+i\sqrt{\pi}(\omega/k\bar{v}_{xj})W(\omega/k\bar{v}_{xj})]\right]A_1/\tilde{\lambda}_j+$$
$$+\delta(y)\sum_j \frac{2V_jc}{\tilde{\lambda}_j\bar{v}_{xj}^2}[1+i\sqrt{\pi}(\omega/k\bar{v}_{xj})W(\omega/k\bar{v}_{xj})]\,\varphi_1, \qquad (2.40)$$

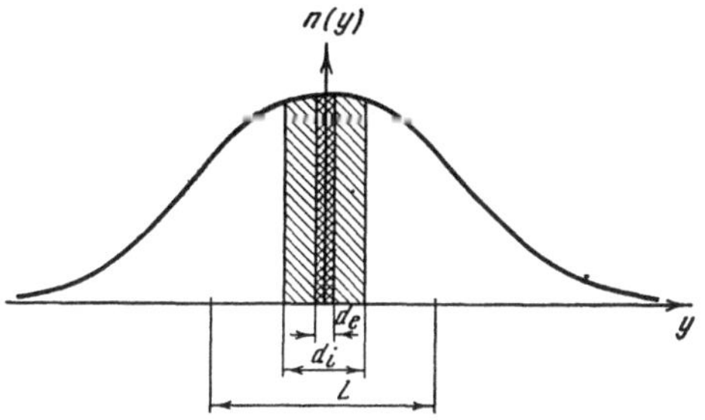

Fig. 2. Structure of the current sheet. The total sheet thickness is L. In the interior regions of thickness d_j, j = e, i, the particles are not magnetized.

$$\left[\frac{d^2}{dy^2} - k^2 + \sum_j \frac{1}{L_{Dj}^2}\right]\varphi_1 = \delta(y)\sum_j \frac{2c^2[1+i\sqrt{\pi}(\omega/k\bar{v}_{xj})W(\omega/k\bar{v}_{xj})]}{\tilde{\lambda}_j \bar{v}_{xj}^2}\varphi_1 - \delta(y)\sum_j \frac{V_j c}{\tilde{\lambda}_j \bar{v}_{xj}^2}[1+i\sqrt{\pi}(\omega/k\bar{v}_{xj})W(\omega/k\bar{v}_{xj})]A_1. \qquad (2.41)$$

Here $L_{Dj} = (\varkappa T_j/4\pi n e_j^2)^{1/2}$ is the Debye length and $\tilde{\lambda}_j = \lambda_j(L/r_{Hj})^{1/2}$.

In the case $|y| = L$, on the other hand, A_1 and φ_1 must satisfy

$$\frac{d^2A_1}{dy^2} - k^2A_1 = 0, \qquad (2.42)$$

$$\frac{d^2\varphi_1}{dy^2} - k^2\varphi_1 = 0. \qquad (2.43)$$

At the boundary $|y| = L$ the logarithmic derivatives of the potentials must be continuous. Equations (2.40)-(2.43) describe perturbations in a current sheet of thickness 2L (Fig. 2).

In the region $|y| < d_j = (r_{Hj}L)^{1/2}$ we can neglect the magnetic field and assume that the particles of species j move along straight paths. The perturbed distribution function in this region is (2.8). Outside the sheet, i.e., at $|y| > d_j$, the plasma is magnetized and [18-20]

$$j_1 = \frac{\partial j_0}{\partial A_0}A_1 + \frac{\partial j_0}{\partial \varphi_0}\varphi_1, \qquad (2.44)$$

$$\rho_1 = \frac{\partial \rho_0}{\partial \varphi_0}\varphi_1 + \frac{\partial \rho_0}{\partial A_0}A_1. \qquad (2.45)$$

The Green's function for boundary-value problem (2.40)-(2.43) is found in the standard manner. The eigenvalue problem for frequency ω is solved by the procedure used above for the thin current sheet. As a result we find the dispersion relations (kL < 1)

$$\left(1 + \sum_j \frac{1 - (V_j^2 + \bar{v}_{zj}^2)/\bar{v}_{xj}^2\,[1+i\sqrt{\pi}(\omega/k\bar{v}_{xj})W(\omega/k\bar{v}_{xj})]}{\tilde{\lambda}_j\sqrt{2/L^2 - k^2}}\right)\left(1 + \sum_j \frac{c^2[1+i\sqrt{\pi}(\omega/k\bar{v}_{xj})W(\omega/k\bar{v}_{xj})]}{\bar{v}_{xj}^2\tilde{\lambda}_j\sqrt{\sum_j 1/L_{Dj}^2 - k^2}}\right) + \left(\sum_j \frac{V_j c[1+i\sqrt{\pi}(\omega/k\bar{v}_{xj})W(\omega/k\bar{v}_{xj})]}{\tilde{\lambda}_j \bar{v}_{xj}^2}\right)^2 \Big/ \sqrt{(2/L^2 - k^2)\left(\sum_j 1/L_{Dj}^2 - k^2\right)} = 0. \qquad (2.46)$$

If $V_i = 0$, $\omega/k\bar{v}_j \ll 1$, then from (2.46) we have

$$\left[1+i\sqrt{\pi}\frac{\omega}{k\bar{v}_e}\left(\left(\frac{L}{r_{H_e}}\right)^{3/2}+\frac{\bar{v}_e}{\bar{v}_i}\left(\frac{L}{r_{H_i}}\right)^{3/2}\right)\right]\left[1+\sum_j\frac{c}{\sqrt{\bar{v}_j V}}\right]+\frac{c}{\bar{v}_e}=0. \qquad (2.47)$$

The growth rate for this instability is

$$\gamma\simeq\frac{k\bar{v}_e}{\sqrt{\pi}}\left(\frac{r_{H_e}}{L}\right)^{3/2}\left(1+\sqrt{\frac{r_{H_i}m}{LM}}\right). \qquad (2.48)$$

In Eqs. (2.38) and (2.39) the sheet thickness is [see also (1.3)]

$$L=\lambda_{0j}(\bar{v}_j/V_j)^2, \qquad (2.49)$$

where

$$\lambda_{0j}=m_j c^2/2\pi n_0 e_j^2. \qquad (2.50)$$

The growth rate in (2.48) was found for the case $L \gg r_{He}, r_{Hi}$. Its dependence on the ion mass is inconsequential. In the limit $L < r_{He}$ [see (2.37)], the growth rate does depend on the ion mass, in contrast with that in (2.48). This result shows that the ion gyroradius is, generally speaking, an important parameter. Returning to [22, 23], we would therefore expect quasi-linear stabilization only if $L \lessdot r_{Hi}$. If this is the case, then with $L > r_{Hi}$ the sheet becomes unstable and collapses over a scale time $\gamma^{-1} \lessdot 1/k\bar{v}_i$.

The behavior of neutral current sheets under laboratory conditions was studied in [6-8]. One way to interpret the results of [7] is to conclude that the neutral current sheet breaks up when the ion gyroradius reaches a level corresponding to the typical growth rate for the instability, on the order of $\gamma \sim k\bar{v}_i$.

9. We turn now to the electromagnetic tearing-mode instability of an unsteady current sheet. As a result of the onset of this instability, over a time $\sim\gamma^{-1}$, where γ is the growth rate of the instability, the current sheet should decay into individual plasma filaments stretched out along the current. Between these filaments, regions form with a low plasma density and with an electric field E parallel to the z axis; particles are accelerated by this field. Let us examine the stability of the accelerated plasma against tearing-mode perturbations.

We assume that at time t = 0 there is a neutral current sheet in vacuum in the y = 0 plane, $n(y) = n_0\delta(y)$, whose particles are accelerated in an external electric field $E_0(t)$ which is parallel to the z axis. We treat the problem which is homogeneous along z and symmetric with respect to y. The particles in the sheet are described by the distribution function

$$f_{0j}(\mathbf{x},t,\mathbf{p})=n_0\delta(y)\,\delta(p_x)\,\delta\left(p_z-p_{0j}+\frac{e_j}{c}A_0(t)\right), \qquad (2.51)$$

where

$$A_0(t)=-c\int^t E_0(t')\,dt', \qquad (2.52)$$

and p_{0j} is the initial momentum of particles of species j in the sheet.

Let us consider perturbations of the electromagnetic field and the plasma particles. The perturbation of the distribution function obeys the equation

$$\frac{\partial f_{1j}}{\partial t}+v_x\frac{\partial f_{1j}}{\partial x}+e_jE_0(t)\frac{\partial f_{1j}}{\partial p_z}=\frac{e_j}{c}\left(\frac{\partial A_1}{\partial t}+v_x\frac{\partial A_1}{\partial x}\right)\frac{\partial f_{0j}}{\partial p_z}-\frac{e_j}{c}\left(v_z\frac{\partial A_1}{\partial x}+c\frac{\partial\varphi_1}{\partial x}\right)\frac{\partial f_{0j}}{\partial p_x}. \qquad (2.53)$$

The characteristics of this equation are

$$p_z-p_z'=\frac{e_j}{c}(A_0(t)-A_0(t')), \qquad (2.54)$$

$$x - x' = cp_x \int_{t'}^{t} dt'' \left[m_j^2 c^2 + p_x^2 + \left(p_z + \frac{e_j}{c} (A_0(t) - A_0(t'')) \right)^2 \right]^{-1/2}. \quad (2.55)$$

The solution of Eq. (2.53) is

$$f_{1j}(\mathbf{x}, t, \mathbf{p}) = \frac{e_j}{c} \int^{t} dt' \left[\left(\frac{\partial A_1'}{\partial t} + v_x' \frac{\partial A_1'}{\partial x} \right) \frac{\partial f_{0j}}{\partial p_z} - \left(v_z' \frac{\partial A_1'}{\partial x} + c \frac{\partial \varphi_1'}{\partial x} \right) \frac{\partial f_{0j}}{\partial p_x} \right]. \quad (2.56)$$

We take the Fourier transform with respect to x of (2.56):

$$f_{1j}(y, k, t) = \int_{-\infty}^{+\infty} f_{1j}(y, x, t) e^{ikx} dx. \quad (2.57)$$

The perturbation of the current density in the sheet is

$$j_1(y, k, t) = \sum_j e_j \int v_z f_{1j} d^2p = \delta(y) \sum_j e_j^2 n_0 c \times$$
$$\times \int \frac{d^2p p_z}{[m_j^2c^2 + p_z^2 + p_x^2]^{1/2}} \left(\frac{\partial f_{0j}}{\partial p_z} \int^{t} dt' \left(\frac{\partial A_1'}{\partial t} - ikv_x' A_1' \right) \exp\left(ik \int_{t'}^{t} v_x(t'') dt'' \right) - \right.$$
$$\left. - ikc \frac{\partial f_{0j}}{\partial p_x} \int^{t} dt' \left\{ \frac{\left[p_z + \frac{e_j}{c} (A_0(t) - A_0(t')) \right] A_1'}{\left[m_j^2 c^2 + p_x^2 + \left[p_z + \frac{e_j}{c} (A_0(t) - A_0(t')) \right]^2 \right]^{1/2}} + \varphi_1' \right\} \exp\left(ik \int_{t'}^{t} v_x(t'') dt'' \right) \right). \quad (2.58)$$

The perturbation of the charge density in the sheet is

$$\rho_1(y, k, t) = \delta(y) \sum_j e_j^2 n_0 \int d^2p \left(\frac{\partial f_{0j}}{\partial p_z} \int^{t} dt' \left(\frac{\partial A_1'}{\partial t} - ikv_x' A_1' \right) \exp\left(ik \int_{t'}^{t} v_x'' dt'' \right) - \right.$$
$$\left. - ikc \int^{t} dt' \left\{ \frac{\left[p_z + \frac{e_j}{c} (A_0(t) - A_0(t')) \right] A_1'}{\left[m_j^2 c^2 + p_x^2 + \left[p_z + \frac{e_j}{c} (A_0(t) + A_0(t')) \right]^2 \right]^{1/2}} + \varphi_1' \right\} \exp\left(ik \int_{t'}^{t} v_x'' dt'' \right) \right). \quad (2.59)$$

Here the prime shows that the function depends on t'; e.g.,

$$v_x' \equiv v_x(t') = cp_x \Big/ \left\{ m_j^2 c^2 + p_x^2 + \left[p_z + \frac{e_j}{c} (A_0(t) - A_0(t')) \right]^2 \right\}^{1/2}. \quad (2.60)$$

Substituting distribution function f_{0j} from (2.51) into (2.58) and (2.59), integrating over the momenta, and substituting (2.58) and (2.59) into the equations for the potentials,

$$\Delta A_1 = -\frac{4\pi}{c} j_1, \quad (2.61)$$

$$\Delta \varphi_1 = -4\pi\rho_1, \quad (2.62)$$

we find the following equations for $A_1(y, k, t)$ and $\varphi_1(y, k, t)$:

$$\frac{d^2A_1}{dy^2} - k^2A_1 = \delta(y)\sum_j\left(\frac{4\pi n_0 e_j^2}{m_j c^2}\left\{\frac{m_j^3c^3A_1}{\left[m_j^2c^2+\left(p_{0j}-\frac{e_j}{c}A_0(t)\right)^2\right]^{3/2}} - k^2c^2\frac{\left(p_{0j}-\frac{e_j}{c}A_0(t)\right)}{\left[m_j^2c^2+\left(p_{0j}-\frac{e_j}{c}A_0(t)\right)^2\right]^{1/2}}\times\right.\right.$$

$$\left.\left.\times\int^t\left[\frac{\left(p_{0j}-\frac{e_j}{c}A_0(t')\right)A_1'}{\left[m_j^2c^2+\left(p_{0j}-\frac{e_j}{c}A_0(t')\right)^2\right]^{1/2}}+\varphi_1'\right]\int_{t'}^{t}\frac{m_jc}{\left[m_j^2c^2+\left(p_{0j}-\frac{e_j}{c}A_0(t'')\right)^2\right]^{1/2}}dt'dt''\right\}\right), \tag{2.63}$$

$$\frac{d^2\varphi_1}{dy^2} - k^2\varphi_1 = \delta(y)\sum_j\frac{4\pi n_0 e_j^2k^2}{m_j}\int^t\left\{A_1'\frac{\left[p_{0j}-\frac{e_j}{c}A_0(t')\right]}{\left[m_j^2c^2+\left(p_{0j}-\frac{e_j}{c}A_0(t')\right)^2\right]^{1/2}}+\varphi_1'\right\}\times$$

$$\times\int_{t'}^{t}\frac{m_jc}{\left[m_j^2c^2+\left(p_{0j}-\frac{e_j}{c}A_0(t'')\right)^2\right]^{1/2}}dt'dt''. \tag{2.64}$$

Here we have discarded the term describing the displacement current; we will examine the validity of this simplification below. We can write a formal solution of Eqs. (2.63) and (2.64) by applying the inverse Helmholtz operator to the right and left sides of these equations. As before, we set $y = 0$; then we find a system of integral equations for $A_1(t, k, y = 0)$ and $\varphi_1(t, k, y = 0)$:

$$A_1(t, k, y=0) = \left(1+\sum_j\frac{m_j^3c^3}{\lambda_{0j}k\left[m_j^2c^2+\left(p_{0j}-\frac{e_j}{c}A_0(t)\right)^2\right]^{3/2}}\right)^{-1}\sum_j\frac{kc^2\left(p_{0j}-\frac{e_j}{c}A_0(t)\right)}{\lambda_{0j}\left[m_j^2c^2+\left(p_{0j}-\frac{e_j}{c}A_0(t)\right)^2\right]^{1/2}}\times$$

$$\times\int^t\left\{A_1'(t', k, y=0)\frac{\left(p_{0j}-\frac{e_j}{c}A_0(t')\right)}{\left[m_j^2c^2+\left(p_{0j}-\frac{e_j}{c}A_0(t')\right)^2\right]^{1/2}}+\varphi_1(t', k, y=0)\right\}\times$$

$$\times\int_{t'}^{t}\frac{m_jc\,dt''}{\left[m_j^2c^2+\left(p_{0j}-\frac{e_j}{c}A_0(t'')\right)^2\right]^{1/2}}dt', \tag{2.65}$$

$$\varphi_1(t, k, y=0) = \sum_j\frac{kc^2}{\lambda_{0j}}\int^t\left\{\frac{\left(p_{0j}-\frac{e_j}{c}A_0(t')\right)A_1(t', k, y=0)}{\left[m_j^2c^2+\left(p_{0j}-\frac{e_j}{c}A_0(t')\right)^2\right]^{1/2}}+\right.$$

$$\left.+\varphi_1(t', k, y=0)\right\}\int_{t'}^{t}\frac{dt''m_jc}{\left[m_j^2c^2+\left(p_{0j}-\frac{e_j}{c}A_0(t'')\right)^2\right]^{1/2}}dt'. \tag{2.66}$$

We define $\mathscr{P}_j(t)$:

$$\mathscr{P}_j(t) = p_{0j} - \frac{e_j}{c}A_0(t). \tag{2.67}$$

1. In the nonrelativistic case

$$\mathscr{P}_j(t) \ll m_j c, \tag{2.68}$$

we have

$$A_1(t)\left(1+\sum_j \frac{1}{\lambda_j k}\right)=\sum_j \frac{k v_j(t)}{\lambda_{0j}} \int^t (t-t')(A_1(t')v_j(t')+c\varphi_1(t'))\,dt', \tag{2.69}$$

$$\varphi_1(t)=\sum_j \frac{kc}{\lambda_{0j}} \int^t (t-t')(A_1(t')v_j(t')+c\varphi_1(t'))\,dt', \tag{2.70}$$

where $v_j(t)=\mathscr{P}_j(t)/m_j$. We assume that the ions are at rest, $v_i = 0$, but that $v_e \neq 0$. Then system (2.69), (2.70) is equivalent to the following system of differential equations:

$$\frac{d^2}{dt^2}\left(\frac{A_1(t)}{v_e(t)}\right)=\frac{k}{\lambda_{0e}\left(1+\sum_j \frac{1}{\lambda_j k}\right)}(A_1(t)v_e(t)+c\varphi_1(t)), \tag{2.71}$$

$$\frac{d^2\varphi_1}{dt^2}=\frac{kc}{\lambda_{0e}}(A_1(t)v_e(t)+c\varphi_1(t))+\frac{kc^2}{\lambda_{0e}}\frac{m}{M}\varphi_1(t). \tag{2.72}$$

We write the time dependence of A_1 and φ_1 in the form $\sim\exp\{-i\int^t \omega(t)dt\}$; then for high velocities $v_e(t)$ we find from (2.71) and (2.72) a dispersion relation like (2.36) for $\omega(t)$. Accordingly, in an unsteady electric field the current sheet is also unstable against the tearing mode. The perturbations increase over time according to

$$\exp\left\{\frac{k\int^t v(t)\,dt}{\sqrt{\lambda_e k+1}}\left(\frac{m}{M}\right)^{1/2}\right\}. \tag{2.73}$$

2. In the ultrarelativistic problem

$$\mathscr{P}_j \gg m_j c, \tag{2.74}$$

we assume $p_{0i} = -p_{0e}$ and $e_e = -e_i$; then we have $\mathscr{P}_i=-\mathscr{P}_e=\mathscr{P}(t)$. The latter equation can also hold if $p_{0i} \neq -p_{0e}$, provided that $eA(t)/c \gg p_{0j}$. In this case we find from Eqs. (2.65) and (2.66) the following system of equations for A_1 and φ_1, where we are neglecting terms $m_j^3c^3/\mathscr{P}^3\lambda_{0j}k$:

$$\frac{d}{dt}\mathscr{P}(t)\frac{dA_1}{dt}=4\pi n_0 e^2 ck(A_1+\varphi_1), \tag{2.75}$$

$$\frac{d}{dt}\mathscr{P}(t)\frac{d\varphi_1}{dt}=4\pi n_0 e^2 ck(\varphi_1+A_1). \tag{2.76}$$

Alternatively,

$$A_1-\varphi_1=\text{const}\,\xi(t), \tag{2.77}$$

$$\frac{d^2(A_1+\varphi_1)}{d\xi^2}-8\pi n_0 e^2 ck\mathscr{P}(\xi)(A_1+\varphi_1)=0. \tag{2.78}$$

Here we have defined

$$\xi(t)=\int^t \frac{dt'}{\mathscr{P}(t')}. \tag{2.79}$$

We assume $\mathscr{P}(t)=\alpha t^n$ and $n \geq 1$; then the solution of Eq. (2.78) is [32]

$$(A_1+\varphi_1)=\frac{\text{const}}{\sqrt{(n-1)\,t^{(n-1)/2}}}\, I_{\frac{n-1}{n-2}}\left(\frac{\sqrt{8\pi n_0 e^2 ck}}{(n-2)\sqrt{\alpha}\, t^{(n-2)/2}}\right). \qquad (2.80)$$

If $n = 1$, in particular, if the external electric field is constant ($\mathscr{P} = p_0 + eE_0t$), then we find the following in the limit $t \to \infty$:

$$A_1+\varphi_1=\frac{\text{const}}{(p+eE_0t)^{1/4}}\exp\left[\frac{\sqrt{8\pi n_0 e^2 ck}}{eE_0}(p_0+eE_0t)^{1/2}\right]\to\infty, \qquad (2.81)$$

$$A_1-\varphi_1=\text{const}\ln(p_0+eE_0t)\to\infty. \qquad (2.82)$$

If $1 < n < 2$, then in the limit $t \to \infty$ we would have

$$A_1-\varphi_1\sim t^{1-n}\to 0, \qquad (2.83)$$

$$A_1+\varphi_1\sim t^{\frac{3n-4}{4}}\exp\frac{\sqrt{8\pi n_0 e^2 ck\alpha^{-1}}\, t^{\frac{(2-n)}{2}}}{(n-2)}, \qquad (2.84)$$

i.e., an instability occurs for $1 \leq n < 2$.

If $n > 2$, then in the limit $t \to \infty$ we have

$$A_1-\varphi_1\sim t^{1-n}\to 0, \qquad (2.85)$$

$$A_1+\varphi_1\sim\frac{1}{(n-1)\,\Gamma(2n-3/n-2)}\left(\frac{8\pi n_0 e^2 ck}{\alpha(n-2)^2}\right)^{(n-1)/2(n-2)} t^{1-n}\to 0. \qquad (2.86)$$

In other words the flux of relativistic particles accelerated by an electric field which increases more rapidly than linearly with the time ($E_0(t) \sim t^m$, $m > 1$) is stable.

Is it legitimate to neglect the displacement current in Eqs. (2.59)? To answer this question we must compare the quantities $c^2k^2A_1$ and d^2A_1/dt^2; from (2.71) and (2.72) we find a ratio $(d^2A_1/dt^2)/A_1c^2k^2 \ll 1$.

3. Simple Models for Charged-Particle Acceleration at Neutral Current Sheets

Let us consider some simple models for the dynamics of a thin current sheet and the surrounding plasma for the case of finite changes in the field or current in the sheet. These changes could be caused, in particular, by tearing-mode instabilities or an external field, as a result of the incidence of an electromagnetic wave of finite amplitude from outside the sheet. In this case strong electric fields $E \sim H$ can arise if the particles in and near the sheet can no longer be treated as a plasma [1, 2, 33], in the sense that a real plasma shields external low-frequency electric and magnetic fields [at frequencies $\omega < \omega_p = (4\pi ne^2/m)^{1/2}$]. This case can arise if the external fields are so strong that their shielding requires a current larger than the limiting current $j_{li} = enc$. In this case the condition for the quasisteady nature of the magnetic fields is not satisfied. In plasmas in space the quasisteady condition is generally satisfied easily, except for plasma motion near null points and neutral current sheets [1, 2].

As before we consider a thin current sheet in the $y = 0$ plane. The sheet is in either a plasma of density n or vacuum. The problem is homogeneous with respect to z and x and symmetric with respect to y. Furthermore, we neglect the magnetic part of the Lorentz force in the equations of motion for the plasma particles outside the sheet; this is legitimate if $\omega_H/\omega_p \ll 1$ or $E \gg H$ [33]. For the particles in the sheet the magnetic part of the Lorentz force drops out because of the symmetry of the problem. Then taking into account conservation of canonical momentum, we find

$$\frac{\partial}{\partial t}\left(p_j + \frac{e_j}{c}A\right) = 0 \tag{3.1}$$

and the Maxwell equations for the field,

$$\frac{\partial^2 A}{\partial y^2} - \frac{1}{c^2}\frac{\partial^2 A}{\partial t^2} = -\frac{4\pi}{c} j(y, t) - \frac{1}{c^2} A(y, t=0)\delta'(t) - \frac{1}{c^2}\frac{\partial A}{\partial t}(y, t=0)\delta(t), \tag{3.2}$$

where the current density is

$$j(y, t) = en(y)c\left[\frac{p_i}{(M^2c^2+p_i^2)^{1/2}} - \frac{p_e}{(m^2c^2+p_e^2)^{1/2}}\right]. \tag{3.3}$$

Here the vector potential is $\mathbf{A} = (0, 0, A)$; p_j is the z-th momentum component of a particle of species j in the plasma, $e_e = -e_i$, $m_e = m$, $m_i = M$, and n(y) is the y dependence of the density of plasma particles.

An exact calculation of the entire unsteady behavior of the plasma and the field runs into serious mathematical difficulties, so we turn to three simple problems in which strong electric fields arise and in which particles are accelerated.

1. We first consider the case of an instantaneous decay of a sheet in a plasma [33]. We assume that the initial (t = 0) magnetic field in a homogeneous plasma with an electron density n is $\mathbf{H} = (H_0 \operatorname{sign} y, 0, 0)$ and that there is no current in the y = 0 plane. Such a situation could arise, for example, upon the rapid decay of a thin current sheet. We assume that the plasma particles are at rest at t = 0. Then the particle momentum p satisfies the equation

$$c^2\frac{\partial^2 p}{\partial y^2} - \frac{\partial^2 p}{\partial t^2} = \omega_p^2\left(\frac{pmc}{\sqrt{m^2c^2+p^2}} + \frac{pmc}{\sqrt{M^2c^2+p^2}}\right) + 2eH_0\delta(y)c, \tag{3.4}$$

where $H_0 = 2\pi I_0/c$.

This equation can be solved easily in two limiting cases. First, in the nonrelativistic limit ($p/mc \ll 1$) the solution is

$$\begin{gathered} E(y, t) = H_0 J_0\left(\omega_p\sqrt{t^2-(y/c)^2}\right)\theta(ct-|y|), \\ H(y, \infty) = H_0[1-\exp(-\omega_p y/c)], \quad p(y, \infty) = -\frac{eH_0}{\omega_p}\exp\left(-\frac{\omega_p y}{c}\right). \end{gathered} \tag{3.5}$$

On both sides of the original discontinuity an electric-field pulse moves at a velocity c; the width of this pulse is inversely proportional to the time. Over a time $t \sim 1/\omega_p$ a current sheet of thickness $l \sim c/\omega_p$ is established at the discontinuity. An energy $\sim H_0^2 c/8\pi\omega_p$ is transferred to electrons. If $H^2/4\pi nmc^2 \gg 1$, we can use the ultrarelativistic limit of Eq. (3.4), i.e., assume $p/mc \gg 1$, for sufficiently early times. Then for the electric field we have a wave equation. In the region $|y| < ct$, instead of the original magnetic field, an electric field of equal magnitude, $E = H_0$, appears, and the particle momentum is

$$p = -eH_0(t - |y|/c) \tag{3.6}$$

for $|y| < ct$. The solution takes this simple form only at times $t < t_m$, where $t_m \ll (\omega_H/\omega_p^2)$ (corresponding to the condition $E \gg H$). The particles are ultrarelativistic at $t_m \gg 1/\omega_H$. These two conditions are compatible if $H^2/4\pi nmc^2 \gg 1$.

2. We now assume that a wave of finite amplitude is incident on the sheet. We assume that there is a thin current sheet in the (x, z) plane in vacuum with a current

$$j_0(y) = I_0\delta(y) = en_0\delta(y)\,c\,[(p_{0i}(M^2c^2 + p_{0i}^2)^{-1/2} - p_{0e}(m^2c^2 + p_{0e}^2)^{-1/2}].$$

A wave of amplitude E_B, described by the potential $A_B(y, t) = -E_B(ct + |y|)\theta(ct + |y|)$ with $t \leq 0$, is incident externally on the sheet. In this case the equation for the field is

$$c^2\frac{\partial^2 a}{\partial y^2} - \frac{\partial^2 a}{\partial t^2} = -2\omega_0 c\delta(y)\left\{\frac{(a+\tilde{p}_e)}{[1+(a+\tilde{p}_e)^2]^{1/2}} + \left(\frac{m}{M}\right)\frac{(a-\tilde{p}_i)}{[1+[(a-\tilde{p}_i)\,m/M]^2]^{1/2}}\right\} + \\ + \frac{2e}{mc^2}(E_B + H_0)\delta'(t)|y| + \frac{2e}{m}E_B\delta(t), \tag{3.7}$$

where $a = eA/mc^2$, $\omega_0 = 2\pi n_0 e^2/mc$, and the initial momenta of the electrons and ions in the sheet are $\tilde{p}_e = p_{0e}/mc$ and $\tilde{p}_i = p_{0i}/mc$.

Equation (3.7) can be rewritten on the basis of the d'Alembert equation, which gives the solution of the Cauchy problem for the wave equation:

$$a(y, t) = -\omega_0\int_0^{t-|y|/c}\left\{\frac{a(y=0, t') + \tilde{p}_e}{[1+(a(y=0, t') + \tilde{p}_e)^2]^{1/2}} + \frac{m}{M}\frac{a(y=0, t') - \tilde{p}_i}{[1+[m/M(a(y=0, t') - \tilde{p}_i)]^2]^{1/2}}\right\}dt' - \\ - \frac{2\pi e I_0}{mc^2}(|y| + \theta(ct - |y|)(ct - |y|)) - \frac{eE_B}{mc^2}\left(\left(t + \frac{y}{c}\right)\theta\left(t + \frac{y}{c}\right) + \left(t - \frac{y}{c}\right)\theta\left(t - \frac{y}{c}\right)\right). \tag{3.8}$$

To determine $a(y = 0, t)$ we set $y = 0$ in (3.8) and differentiate with respect to t; the result is an ordinary first-order differential equation, whose solution in this case is

$$\int_0^{a(0,t)}\left\{\frac{(a' + \tilde{p}_e)}{\sqrt{1+(a'+\tilde{p}_e)^2}} + \frac{m}{M}\frac{(a' - \tilde{p}_i)}{\sqrt{1+[(a'-\tilde{p}_i)m/M]^2}} + \left(\frac{2\pi e I_0}{mc^2\omega_0} + \frac{2eE_B}{mc\omega_0}\right)\right\}^{-1} da' = -\omega_0 t. \tag{3.9}$$

We see immediately from (3.9) that if $E_B > \pi(2en_0c - I_0)/c$ there can be no steady-state solution:

$$a(y=0, t) \sim -e\left(E_B - \frac{\pi}{c}(2en_0c - I_0)\right)t/mc \quad \text{as} \quad t \to \infty \tag{3.10}$$

Consequently [see (3.1)], the particle momentum in the sheet increases without bound,

$$p(y=0, t\to\infty) \sim e\left(E_B - \frac{\pi}{c}(2en_0c - I_0)\right)t. \tag{3.11}$$

Finding $a(y = 0, t)$ from (3.9) and substituting it into (3.8) we can find equations for the electromagnetic field outside the sheet. The solution for $E(y, t)$ far behind the wave front consists of two "steps" of height $E_B - \pi(2en_0c - I_0)/c$, extending in both directions. If the amplitude of the wave incident on the sheet is not very high, $E_B < \pi(2en_0c - I_0)/c$, the wave is reflected from the neutral sheet. The particle momentum in the sheet increases. In the region $|y| < ct$ the electric field is $E = 0$ and the magnetic field is $H(y) = (H_0 + E_B)\,\mathrm{sign}\,y$.

3. We now assume that at time $t = -0$ there is a current sheet in the $y = 0$ plane with a current $I_0 = 2en_0V$; the magnetic field produced by this current is $\mathbf{H} = H_0(\mathrm{sign}\,y, 0, 0)$. At

time t = +0 the particle density in the sheet is n_1 so that the current is $I(t = +0) \equiv I_1 = 2en_1V$. Such a situation could presumably arise, for example, in the case of an incomplete decay of a thin current sheet, in which case the particle density in the sheet drops, not to zero, but to some finite value. In this case the solution of system (3.1)-(3.3) can be found by the procedure leading to (3.9); this solution is

$$\int_0^{a(y=0,t)} da' \left\{\frac{2\pi e I_0}{mc^3\omega_1} + \frac{(a'+\bar{p}_e)}{\sqrt{1+(a'+\bar{p}_e)^2}} + \frac{m}{M}\frac{(a'+\bar{p}_i)}{\sqrt{1+\left[\frac{m}{M}(a'+\bar{p}_i)\right]^2}}\right\}^{-1} = \omega_1 t. \tag{3.12}$$

Here $\omega_1 = 2\pi e^2 n_1/mc$ and the notation is otherwise the same as above.

Again in this case there is no steady-state solution if the perturbation is large. If $I_0 > 2en_1c$, i.e., if $n_1 < n_0V/c$, there is an unbounded acceleration of the particles in the sheet; in the limit $t \to \infty$ the momentum of these particles is described by

$$p(t \to \infty) \sim \frac{2\pi e}{c}(I_0 - I_1)t = \frac{4\pi e^2}{c}(n_0V - n_1c)t. \tag{3.13}$$

We thus see that effective particle acceleration is possible in principle near neutral current sheets; in one case a large part of the magnetic energy is transferred to the particles, and in other cases these particles are accelerated to essentially unlimited energies. For this acceleration to occur, however, several conditions must be satisfied; for example, the sheet must be homogeneous, the breakup of the sheet must occur rapidly ("instantaneously") at a certain stage of its development, it must be possible for relativistic current sheets to form, and, finally, it must be legitimate to neglect the influence on the particle acceleration of the streaming instabilities which should appear in such a system. How well these conditions can be satisifed in a real plasma requires further study.

4. Model for the Decay of a Current Sheet

We have now examined the particle acceleration in current sheets under conditions such that the plasma density in the sheet decreases abruptly in a very short time interval (Section 3). In this case an electric field $E \sim H_0$ arises. Since we were dealing with systems homogeneous along the sheet this field was also homogeneous. Such a decrease in the plasma density in a sheet can occur as the result of the onset of the electromagnetic tearing-mode instability. In our analysis of this instability (Section 2) we dealt with only the beginning of the sheet breakup. The situation which results is far from homogeneous: The current sheet breaks up into plasma filaments $\lambda \sim 2\pi L$ in size, where L is the sheet thickness. The corresponding regions in which the electric fields appear are of the same size. Accordingly, it seems worthwhile to find the electromagnetic fields which arise in a system which is inhomogeneous along x. We examine a model for the decay of a current sheet below, assuming the time evolution of the sheet to be given. Specifically, we assume that at time t = 0 there is a thin current sheet in the y = 0 plane which is homogeneous along x and that the magnetic field is $\mathbf{H} = 2\pi I_0/c \times$ (sign y, 0, 0). At t > 0 the time and coordinate dependences of the current density are

$$j_z(x, y, t) = I_0(1 - \theta(x(t) - |x|))\delta(y). \tag{4.1}$$

In other words the current sheet breaks into halves, which move along the x axis in accordance with a given law $x = \pm x(t)$.

For the electromagnetic potential $A_z(x, y, t)$ we have the wave equation

$$\frac{\partial^2 A_z}{\partial x^2} + \frac{\partial^2 A_z}{\partial y^2} - \frac{1}{c^2}\frac{\partial^2 A_z}{\partial t^2} = -\frac{4\pi}{c}j_z \tag{4.2}$$

with the initial conditions

$$A_z(t=0) = -\frac{2\pi I_0}{c}|y|, \tag{4.3}$$

$$\frac{\partial A_z}{\partial t}(t=0) = 0. \tag{4.4}$$

Solving problem (4.2)-(4.4) with current (4.1), we find the electric field $E_z = -c^{-1}\partial A_z/\partial t$ to be

$$E_z(x, y, t) = \frac{2I_0}{c^2}\int_0^t \left\{\frac{\theta(t-\tau-\sqrt{(x-x(\tau))^2+y^2}/c)}{[(t-\tau)^2-((x-x(\tau)^2+y^2)/c^2)]^{1/2}} + \frac{\theta(t-\tau-\sqrt{(x+x(\tau))^2+y^2}/c)}{[(t-\tau)^2-((x+x(\tau))^2+y^2)/c^2]^{1/2}}\right\}\dot{x}(\tau)\,d\tau. \tag{4.5}$$

The magnetic-field component $H_y(x, y, t)$ is

$$H_y(x, y, t) = \frac{2I_0}{c^2}\int_0^t \left\{\frac{\theta(t-\tau-\sqrt{(x+x(\tau))^2+y^2}/c)}{[(t-\tau)^2-((x+x(\tau))^2+y^2)/c^2]^{1/2}} - \frac{\theta(t-\tau-\sqrt{(x-x(\tau))^2+y^2}/c)}{[(t-\tau)^2-((x-x(\tau))^2+y^2)/c^2]^{1/2}}\right\}d\tau. \tag{4.6}$$

The component $H_x(x, y, t)$ vanishes at the current sheet ($y = 0$) by virtue of the symmetry of the problem. If the coordinate x of the boundary of a gap in the sheet is a linear function of the time,

$$x(t) = \pm vt, \tag{4.7}$$

the integrals in (4.5) and (4.6) can be evaluated easily. The electric field turns out to be

$$E_z(x, y, t) = \frac{I_0 v}{2c^2\sqrt{1-v^2/c^2}}\ln\left[\frac{(t_+ + \sqrt{t^2-r^2/c^2})(t_- + \sqrt{t^2-r^2/c^2})}{(t_+ - \sqrt{t^2-r^2/c^2})(t_- - \sqrt{t^2-r^2/c^2})}\right], \tag{4.8}$$

where

$$t_\pm = \frac{t \pm xv/c}{\sqrt{1-v^2/c^2}} \tag{4.9}$$

and $r^2 = x^2 + y^2$. The magnetic-field component $H_y(x, y, t)$ is

$$H_y(x, y, t) = \frac{I_0}{2c\sqrt{1-v^2/c^2}}\ln\left[\frac{(t_+ + \sqrt{t^2-r^2/c^2})(t_- - \sqrt{t^2-r^2/c^2})}{(t_+ - \sqrt{t^2-r^2/c^2})(t_- + \sqrt{t^2-r^2/c^2})}\right]. \tag{4.10}$$

At the point $\mathbf{r} = 0$ the magnetic field is $\mathbf{H} = 0$ for all t. The electric field is constant, given by

$$E_z(\mathbf{r}=0, t) = \frac{I_0 v}{c^2\sqrt{1-v^2/c^2}}\ln\left[\frac{1+\sqrt{1-v^2/c^2}}{1-\sqrt{1-v^2/c^2}}\right] = \frac{H_0}{2\pi}\frac{v/c}{\sqrt{1-v^2/c^2}}\ln\left[\frac{1+\sqrt{1-v^2/c^2}}{1-\sqrt{1-v^2/c^2}}\right]. \tag{4.11}$$

Here H_0 is the magnetic field outside the sheet at $t = 0$. For small velocities $v/c \ll 1$, we have

$$E(0, t) \simeq H_0\frac{v}{\pi c}\ln\left(\frac{c}{2v}\right), \tag{4.12}$$

i.e., as $v/c \to 0$ the electric field at the rupture tends toward zero more slowly than linearly.

In the ultrarelativistic case, with $v/c \simeq 1$, we have

$$E(0, t) \simeq \frac{v}{\pi c}H_0. \tag{4.13}$$

We thus see that electric fields $E \sim H_0 v/c$ are generated as the current sheet decays; here H_0 is the field outside the sheet and v is the characteristic plasma velocity.

It follows from Eqs. (4.5) and (4.6) and the solution of problem (4.2)-(4.4) for the magnetic-field component H that near the point r = 0 ($|r| \ll vt$) the vector potential takes the following form in the limit $t \to \infty$:

$$A(\mathbf{r}, t) \approx \frac{1}{\pi} H_0 vt - \frac{H_0}{\pi vt}(x^2 - y^2), \tag{4.14}$$

i.e., a singular magnetic null point of the hyperbolic type appears at the rupture [4]. The electric field is constant, while the gradient of the magnetic field falls off as time elapses.

Literature Cited

1. S. I. Syrovatskii, Astron. Zh., 43:349 (1966).
2. V. S. Imshennik and S. I. Syrovatskii, Zh. Éksp. Teor. Fiz., 52:990 (1967).
3. S. I. Syrovatskii, Tr. FIAN, 74:3 (1974).
4. S. I. Syrovatskii, Zh. Éksp. Teor. Fiz., 60:1727 (1971).
5. B. V. Somov and S. I. Syrovatskii, Zh. Éksp. Teor. Fiz., 61:1864 (1971).
6. N. Ohyabu and N. Kawashima, J. Phys. Soc. Jpn., 33:496 (1972).
7. M. Alidières, R. Aymar, P. Jourdan, F. Koechlin, and A. Samain, Plasma Phys., 10:841 (1968).
8. S. I. Syrovatskii, A. G. Frank, and A. Z. Khodzhaev, ZhÉTF Pis. Red., 15:138 (1972).
9. S. I. Syrovatskii, in: Solar Flares and Space Research (ed. Z. Svestka and C. de Jager), North-Holland, Amsterdam (1969), p. 346.
10. S. I. Syrovatskii, in: Solar-Terrestrial Physics, 1970 (E. R. Dyer, ed.), D. Reidel, Dordrecht (1972), p. 119.
11. E. G. Harris, Nuovo Cimento, 23:115 (1962).
12. S. I. Syrovatskii, in: Critical Problems of Magnetospheric Physics. Proceedings of the Symposium, 1972, IUSTP, Washington (1972), p. 35.
13. V. K. Neil, Phys. Fluids, 5:14 (1962).
14. H. P. Furth, Nucl. Fusion, Suppl. Part 1, 169 (1962).
15. H. P. Furth, I. Killeen, and M. N. Rosenbluth, Phys. Fluids, 6:459 (1963).
16. D. Pfirsch, Z. Naturforsch., 17A:861 (1962).
17. G. Laval, R. Pellat, and M. Vuillemin, Proceedings of the Conference on Plasma Physics and Controlled Nuclear Fusion, Culham, 1965, Vol. II, Vienna (1966), p. 259.
18. B. Coppi, G. Laval, and R. Pellat, Phys. Rev. Letters, 16:1207 (1966).
19. F. C. Hoh, Phys. Fluids, 9:277 (1966).
20. M. Dobrowolny, Nuovo Cimento, 55B:427 (1968).
21. K. Schindler and M. Soop, Phys. Fluids, 11:1192 (1968).

21a. K. Schindler and M. Soop, in: Critical Problems of Magnetospheric Physics. Proceedings of the COSPAR Symposium, 1972, IUSTP, Washington (1972), p. 52.

22. D. Biskamp, R. Z. Sagdeev, and K. Schindler, Cosmic Electrodynam., 1:297 (1970).
23. O. A. Pokhotelov, Geomagnetizm i Aéronomiya, 12:693 (1972).
24. T. Dzh. Dollan, Geomagnetizm i Aéronomiya, 12:18 (1972).
25. V. N. Faddeeva and N. M. Terent'ev, Tables of the Probability integral of Complex Argument [in Russian], Gostekhizdat, Moscow (1954).
26. B. D. Fried and S. D. Conte, The Plasma Dispersion Function, Academic Press, New York (1961).
27. A. B. Mikhailovskii, Theory of Plasma Instabilities, Consultants Bureau, New York (1973).

28. A. B. Mikhailovskii, in: Reviews of Plasma Physics, Vol. 6, Consultants Bureau, New York (1974).
29. F. Stern, Phys. Rev. Letters, 18:546 (1967).
30. V. L. Ginzburg and A. A. Rukhadze, Waves in a Magnetized Plasma [in Russian], Nauka Moscow (1970).
31. V. L. Ginzburg, Propagation of Electromagnetic Waves in a Magnetized Plasma [in Russian], Nauka, Moscow (1967).
32. E. Jahnke, F. Emde, and F. Lösch, Tables of Higher Functions, McGraw-Hill, New York (1960).
33. S. I. Syrovatskii, Izv. Akad. Nauk SSSR, Ser. Fiz., 31:1303 (1967).
34. P. J. Baum, A. Bratenahl, M. Kao, and R. S. White, "Plasma instability at an X-type magnetic neutral point," Phys. Fluids, 16:1501 (1973).
35. M. Friedman and S. M. Hamberger, Astrophys. J., 152:667 (1968); Solar Phys., 8:104 (1969).

EXPERIMENTAL STUDY OF THE CONDITIONS FOR THE APPEARANCE OF A NEUTRAL CURRENT SHEET IN A PLASMA: SOME CHARACTERISTICS OF THE SHEET

A. G. Frank

Early experiments on the conditions for the formation of a neutral current sheet near a magnetic null line showed that the anomalously high plasma resistance caused by the onset of small-scale instabilities prevents the formation of a neutral sheet. However, the turbulent conductivity can be increased by increasing the plasma density; a method is described for producing a dense plasma in a quadrupole field. A neutral current sheet develops in a plasma when the plasma conductivity reaches $\sim 10^{14}$ esu. A neutral sheet has been produced with a width five times its thickness. The magnetic field at the sheet boundary is six times its initial value; i.e., there is a high concentration of magnetic energy in the current sheet.

INTRODUCTION

The events which occur as a plasma moves near a null line of a magnetic field are attracting much interest in connection with the problem of the collisionless dissipation of magnetic energy and particle acceleration in a plasma. Historically, the interest in these questions arose in attempts to explain the series of events which occur during a solar flare. A flare is a complicated transient process accompanied by rapid variations and dissipation of magnetic fields, intense radiation in the optical, x-ray, and radio ranges, and the ejection of plasma streams and the generation of high-energy particles (solar cosmic rays). Giovanelli [1] was the first to raise the possibility of a relationship between solar flares and magnetic null lines; many subsequent observations were carried out by Severnyi [2]. Null lines are a very common phenomenon in space, especially where there are rapid movements of macroscopic plasma formations [3].

Astrophysical observations have contributed to the development of the theory of plasma dynamics in nonuniform magnetic fields with null lines and have aided the search for effective particle-acceleration mechanisms (see [4, 5] and the literature cited there). It is highly probable that solar flares are an example of some universal mechanism by which magnetic energy is converted into plasma energy and into energy of accelerated particles, so that research in this field is of general physical importance.

Magnetic null lines are particularly important because strong electric fields and high currents can exist in a magnetized plasma only near such null lines. Accordingly, near null lines we could find a unique type of magnetic-energy accumulation, followed by a rapid liberation of this energy and its transfer to the plasma and to high-energy particles [6]. Although the special role played by null lines was understood a comparatively long time ago, the detailed nature of the corresponding plasma flow and the energy-conversion mechanisms remained puzzling for a long time, because serious theoretical difficulties obstruct analysis of plasma motion in a nonuniform magnetic field with a null line, even in the MHD approach. Understandably, the early papers on these questions were extremely crude. On the basis of a qualitative analysis of the equilibrium state of a plasma near a null line, Dungey [7] concluded that such a state is unstable. Dungey was the first to point out the transient nature of the processes occurring near a magnetic null line. In contrast, steady-state models of the plasma flow near null lines were examined in [8-10], and answers were found to the questions of how fast magnetic energy is converted into the energy of a heated plasma due to the finite conductivity [8, 9] and how fast energy is converted at the front of slow-mode, steady-state shock waves [10]. Friedman and Hamberger [11, 12] adopted Petschek's model [10] to examine the influence of the turbulent plasma conductivity on magnetic-energy dissipation.

The MHD stage of the process was analyzed theoretically in a more systematic manner by Syrovatskii and Imshennik [13-17], who showed that the motion of an ideally conducting magnetized plasma in a magnetic field with a null line should lead to the formation of a neutral current sheet, i.e., a sheet which separates regions in which the magnetic fields are equal in magnitude but opposite in direction, with lines of force stretched out along the sheet. The formation of a neutral current sheet near a magnetic null line implies an important increase in the magnetic-energy density in this region. The plasma flow which tends to form a current sheet is of the nature of an accumulation, and the plasma energy increases as external energy is supplied.

Study of the hydrodynamic plasma flow [18, 19] revealed a progressive decrease in the density everywhere near a current sheet. In principle the density can fall to an arbitrarily low value near a sheet, and the sheet itself can become macroscopically unstable. Under these conditions strong pulsed electric fields can arise and can accelerate charged particles [20]. Such processes, which are typical in astrophysical situations, are very probably operating in the earth's magnetosphere, in solar flares, and in the production of cosmic rays at their remote sources.

It is possible that such events can also be observed in a variety of experiments with laboratory plasmas, specifically, in the fast processes during changes in the magnetic field configuration in a Z pinch and, especially, in a Θ pinch [21, 22]. The acceleration mechanisms involved in these processes have not yet been finally resolved.

Another reason for studying the behavior of plasmas near null lines is the possibility of thereby developing a new plasma acceleration method: It may become possible to produce strong pulsed electric fields by producing a neutral current sheet with a high energy concentration (the energy is in the form of the magnetic energy of the plasma current) and exploiting the rapid rupture of this sheet [23].

The conclusion that a high energy density could be produced in the magnetic field of the current in a neutral sheet was reached on the basis of an MHD analysis of the motion of an ideal plasma. Since the several simplifying assumptions used in this analysis may be quite at odds with the actual situation, experiments are crucial. Only a few pertinent experimental studies have been reported. Bratenahl and Yeates [24] produced a magnetic null line by arranging the convergence of two shock waves, and they studied the dissipation of the magnetic fields carried by the shock waves. They believe that their experiment is in qualitative correspondence with the model of [10]. Ohyabu and Kawashima [25] recently reported the formation of a current sheet as a current flows along a null line of a magnetic field with a special configuration, produced by external conductors. Photographs of the glow of the current sheets reveal ruptures in it and reveal plasma ejections along the magnetic lines of force of the current sheet.

Below we report experiments carried out to study the formation of a neutral current sheet in the case of two-dimensional plasma flow in a magnetic field with a null line and to study the possibility of producing a high concentration of current and magnetic energy near the null line. This work was spurred by the theory of [13-17] and represents the first step in the study of the possible production of strong pulsed electric fields during the collapse of current sheets under laboratory conditions.

CHAPTER I

EVENTS WHICH OCCUR AS A PLASMA MOVES IN A MAGNETIC FIELD WITH A NULL LINE

§ 1. Magnetic Null Lines

"Null lines" or "neutral lines" of a magnetic field are lines at which the magnetic field vector **H** vanishes. Let us construct a Cartesian coordinate system at some point on a null line with Oz axis directed along the null line. Near this point the magnetic field has no H_z component, within terms of higher order in the powers of the coordinates, and its lines of force lie in the (x, y) plane, which is perpendicular to the null line. Such two-dimensional fields can be described [26] by means of a vector potential **A** which has only a z component:

$$\mathbf{A} = \{0;\ 0;\ A(x,\ y,\ t)\}. \tag{1}$$

In this case the vector **H** is

$$\mathbf{H} = \operatorname{rot}\mathbf{A} = \operatorname{rot} A\mathbf{e}_z = [\nabla A \times \mathbf{e}_z] = \left\{\frac{\partial A}{\partial y};\ -\frac{\partial A}{\partial x};\ 0\right\}. \tag{2}$$

According to Eq. (2) and the definition of a line of force [27], tne lines of force in the z = const plane at time t are the families of lines

$$A(x,\ y,\ t) = \text{const} \quad \text{for} \quad t = \text{const}, \quad z = \text{const}. \tag{3}$$

The magnetic flux between lines of force A_1 and A_2 per unit distance along the z axis is

$$\Phi = \int_{A_1}^{A_2} \mathbf{H}\,[\mathbf{e}_z d\mathbf{l}] = -\int_{A_1}^{A_2} \nabla A\, d\mathbf{l} = A_1 - A_2; \tag{4}$$

i.e., the difference between the values of A at two lines of force is equal to the magnetic flux between them.

In general the magnetic field near a null line can be quite complicated, but its absolute value $|\mathbf{H}|$ must increase in proportion to r^n with distance from the null line; here $r = (x^2 + y^2)^{1/2}$ is the distance from the observation point to the null line. The exponent n is called the "order of the null line."

The magnetic field near the simplest type of null line, a first-order null line (Fig. 1), is described by the vector potential

$$A = \frac{h_0}{2}(x^2 - y^2) + C_0, \tag{5}$$

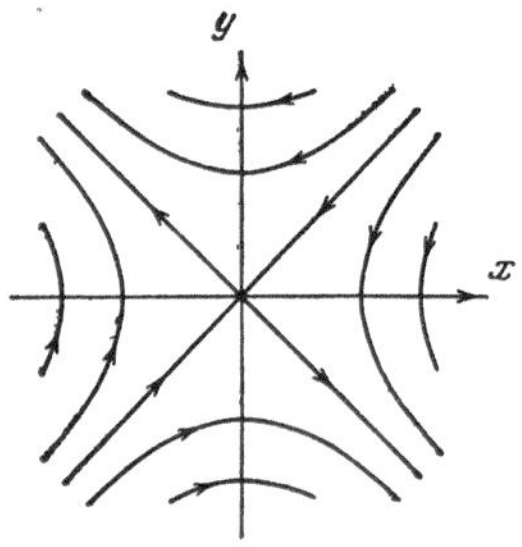

Fig. 1. Magnetic lines of force near a first-order magnetic null line (x = 0, y = 0) in the plane perpendicular to the null line.

where h_0 and C_0 are constants independent of the coordinates. According to Eqs. (2) and (5) the magnetic field in this case is

$$\mathbf{H}=\{-h_0y;\ -h_0x;\ 0\},\qquad (6)$$
$$|\mathbf{H}|=h_0\sqrt{x^2+y^2}=h_0r,$$

increasing linearly with r; the constant h_0 is equal to the absolute value of the gradient in the magnetic field.

§ 2. Steady-State Models for the Conversion of Magnetic Energy into Plasma Energy near Null Lines

Several theoretical papers [8-10] have dealt with steady-state models for plasma flow near null lines, and the rate at which magnetic energy is converted into plasma energy has been determined. An attempt was made to explain the events observed during solar flares. The model of [8] involves a one-dimensional steady-state plasma flow with a frozen-in magnetic field on the two sides of a plane boundary between two regions in which the magnetic fields are oppositely directed (Fig. 2). A strictly one-dimensional situation cannot exist in the steady state, so it is also necessary to take into account the outward flow of plasma along the boundary between the different field regions. A null point is assumed to exist within the boundary, and it is assumed that near this point, because of the finite conductivity, there is a dissipation of the magnetic field carried by the plasma flow. The situation is maintained in a steady state because the velocity at which the field (and the plasma) enter the boundary layer, u_{y_0}, is equal to the velocity of the magnetic-field diffusion with respect to the plasma due to the finite conductivity [9]:

$$u_{y_0}=\frac{c^2}{4\pi\sigma\delta},\qquad (7)$$

where 2δ is the thickness of the boundary layer in which the magnetic field is dissipated (Fig. 2), σ is the plasma conductivity, and c is the speed of light. However, the magnetic en-

Fig. 2. Steady-state model for magnetic field dissipation in the case of two-dimensional plasma flow [8, 9].

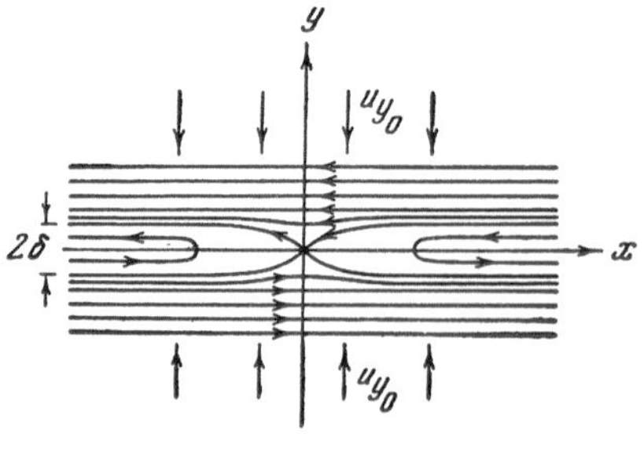

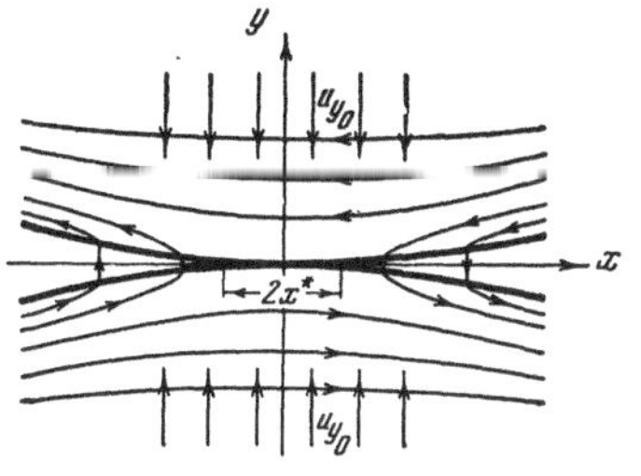

Fig. 3. Steady-state model for magnetic-field dissipation with slow-mode oblique shock waves taken into account [10].

ergy is not converted rapidly enough in this model: Estimates based on this model yield times which are longer than those observed for solar flares by a factor of ~10-100.

A much faster dissipation of magnetic energy can be obtained on the basis of Petschek's model [10], in which it is assumed that slow-mode oblique shock waves exist at a certain distance from a null line (Fig. 3). The situation is maintained in a steady state because the propagation velocity of the slow-mode shock wave with respect to the plasma is equal to the plasma flow velocity opposite the wave u_{y0}. At the shock front, magnetic energy is converted into plasma thermal energy; the rate of this conversion, u_{y0}, can be much higher than the rate of the ohmic diffusion of the magnetic field, (7), and is essentially independent of the plasma conductivity. The wave velocity is governed by the component of the magnetic field normal to the wave front. Near the null line the magnetic field has essentially no normal component, and in this region, as in [8, 9], the energy conversion is governed by ohmic diffusion. The length of the diffusion region is

$$x^* = \frac{c^2}{8\pi\sigma v_A M_0}, \quad (8)$$

where $V_A = H_x/(4\pi\rho)^{1/2}$ is the Alfvén velocity, H_x is the component of the magnetic field parallel to the boundary separating the oppositely directed magnetic field, ρ is the plasma density, $M_0 = u_{y0}/v_A$, and u_{y0} is the flow velocity of the plasma in which the magnetic field is frozen; here $M_0 \lessapprox 0.1$.

In the case $x > x^*$, processes related to wave propagation dominate. The shock fronts are slightly inclined with respect to the boundary layer (with respect to the x axis) and make an angle $\varphi \simeq M_0 = u_{y0}/v_A$ with each other. The electric current j_z at the front provides an abrupt change in the magnetic field beyond the front. The ratio of the normal component of the magnetic field, H_y, to the tangential component in the incoming plasma flow, H_x, is equal to the ratio of the flow velocity to the Alfvén velocity: $H_y/H_x \simeq M_0 = u_{y0}/v_A$.

According to the Petschek model [10], there should be a normal component of the magnetic field (H_y) in the wave-propagation region, $x > x^*$, and the current-density profile over y (for $x > x^*$) should be described by a two-humped curve with a minimum at $y = 0$. With distance from the null line along the x axis the distance between the two maxima on the $j_z(y)$ curve should increase. It should be noted that the condition for the existence of such flows and the very possibility of their appearance were not discussed in [10]; the matter was left completely unresolved. Nevertheless, Petschek's model for magnetic energy dissipation near a null line is one of the most widely adopted models, especially outside the Soviet Union. It is true that certain investigators have expressed doubt regarding the possible appearance of such flows [4], and it has been pointed out that the this model runs into certain theoretical difficulties [28]. One of the results of the present work (see Chapter V) is the conclusion that Petschek's model [10] does not hold up experimentally.

§ 3. Plasma Motion near a Magnetic Null Line in the MHD Approximation

The general two-dimensional problem of plasma motion in a nonuniform magnetic field containing a null line was formulated in [13, 14]. The motion arose from a given initial state with a change in the field sources and was analyzed on the basis of the system of MHD equations.

We assume that Oz axis of a Cartesian coordinate system is directed along the magnetic null line (Fig. 1). We consider two-dimensional motions for which the velocity $\mathbf{v} = \{v_x(x, y, t); v_y(x, y, t); 0\}$ and the magnetic field $\mathbf{H} = \{H_x(x, y, t); H_y(x, y, t); 0\}$ lie in the (x, y) plane and are independent of the coordinate z. Then, neglecting the dissipative terms, we write the MHD equations as

$$\frac{d\rho}{dt} \equiv \frac{\partial\rho}{\partial t} + (\mathbf{v}\nabla)\rho = -\rho \operatorname{div} \mathbf{v}, \tag{9}$$

$$\frac{dA}{dt} \equiv \frac{\partial A}{\partial t} + (\mathbf{v}\nabla) A = 0, \tag{10}$$

$$\frac{d\mathbf{v}}{dt} = -\frac{1}{\rho}\nabla p - \frac{1}{4\pi\rho}\Delta A \nabla A. \tag{11}$$

Here we have used Eqs. (1)-(4) and the equation for the current density j in terms of the magnetic vector potential:

$$\mathbf{j} = \frac{c}{4\pi} \operatorname{rot} \mathbf{H} = -\frac{c}{4\pi} \Delta A \mathbf{e}_z. \tag{12}$$

Equations (9) and (11) have obvious meanings (they are the equations of continuity and motion), while Eq. (10) states that the magnetic field is frozen in the plasma, so that the magnetic lines of force are carried along with the plasma as it moves.

As the original equilibrium state we consider a current-free plasma at rest, with a constant density and a constant pressure:

$$\mathbf{v}_0 = 0, \quad \Delta A_0 = 0, \quad \rho_0 = \text{const}, \quad p_0 = \text{const}. \tag{13}$$

In a study of hydrodynamic flows in [13, 14] it was assumed that magnetic forces dominate the plasma dynamics; i.e., use was made of the approximation of a strong magnetic field and a cold plasma:

$$v_A \gg v_s, \tag{14}$$

where $V_A = H/(4\pi\rho_0)^{1/2} = |\nabla A_0|/(4\pi\rho_0)^{1/2}$ is the Alfvén velocity and $v_s = (\gamma p_0/\rho_0)^{1/2}$ is the sound velocity. Condition (14) was assumed to hold everywhere except in a small neighborhood of the null line, within a characteristic radius r_s, i.e., this condition was assumed to hold for $r > r_s$. In the case of a first-order null line this characteristic radius is

$$r_s = \frac{\sqrt{4\pi\gamma p_0}}{h_0}. \tag{15}$$

The strong-field condition in (14) allows us to neglect the pressure gradient in Eq. (11); the plasma acceleration must be directed normal to the magnetic lines of force.

The most interesting effects near a null line arise when the plasma motion is caused by an electric field directed along the null line,

$$\mathbf{E} = -\frac{1}{c}\frac{\partial A}{\partial t}\mathbf{e}_z, \tag{16}$$

i.e., in the case in which the vector potential $\mathbf{A}_0$ acquires a time-dependent increment $\beta(t)$:

$$\beta(t) = c\int_0^t E(\tau)\,d\tau, \qquad A = A_0 - \beta(t). \tag{17}$$

Let us examine the nature of the plasma motion in a magnetic field with a null line in the approximation of small perturbations [6, 14]. We write the vector potential and the magnetic field near a first-order null line in the following manner in a cylindrical coordinate system:

$$A_0\{r, \varphi, 0\} = \frac{h_0}{2} r^2 \cos 2\varphi + C_0, \tag{5'}$$

$$\mathbf{H}_0 = \{H_r;\ H_\varphi;\ 0\} = \left\{\frac{1}{r}\frac{\partial A}{\partial \varphi};\ -\frac{\partial A}{\partial r};\ 0\right\} = -h_0 r\{\sin 2\varphi;\ \cos 2\varphi;\ 0\}. \tag{6'}$$

If the potential behavior $\beta_1(t)$ is specified at some circle $r = R$, then in the region $r \leq R$ the potential perturbation $A_1(r, \varphi, t)$ corresponds to a converging cylindrical wave:

$$A(r, \varphi, t) \equiv A_0 + A_1 = \frac{h_0}{2}\left[r^2 \cos 2\varphi - \beta_1\left(t + \frac{\sqrt{4\pi\rho_0}}{h_0}\ln\frac{r}{R}\right)\right] + C_0. \tag{18}$$

The wave propagation velocity decreases as the null line is approached,

$$v(r) \equiv -\frac{dr}{dt} = -\frac{h_0 r}{\sqrt{4\pi\rho_0}} = -v_A, \tag{19}$$

and a potential perturbation propagates through the plasma at the local Alfvén velocity v_A. After the wave front has passed, i.e., at $t \gg [(4\pi\rho_0)^{1/2}/h_0]\ln R/r$, the vector potential becomes that in (17).

The magnetic field in the propagating wave, the current density along the null line, and the plasma velocity and density are

$$H_r = -h_0 r \sin 2\varphi, \tag{20}$$

$$H_\varphi = -h_0 r \cos 2\varphi + \sqrt{\pi\rho_0}\,\frac{\dot{\beta}_1}{r}, \tag{21}$$

$$j_z = \frac{c\rho_0}{2h_0}\frac{\ddot{\beta}_1}{r^2}, \tag{22}$$

$$v_r = \frac{\dot{\beta}_1}{2r}\cos 2\varphi, \tag{23}$$

$$v_\varphi = -\frac{\dot{\beta}_1}{2r}\sin 2\varphi, \tag{24}$$

$$\rho = \rho_0\left[1 + \frac{\cos 2\varphi}{r^2}\left(\beta_1 - \frac{\sqrt{\pi\rho_0}}{h_0}\dot{\beta}_1\right)\right]. \tag{25}$$

Here $\dot{\beta}_1 = d\beta_1/dt$ and $\ddot{\beta}_1 = d^2\beta_1/dt^2$.

Solution (18), (20)-(25) was obtained in the linear approximation; i.e., it holds at sufficiently large distances from the null line:

$$r^2 \gg \beta_1, \qquad r^2 \gg \frac{\sqrt{\pi\rho_0}}{h_0}\dot{\beta}_1. \tag{26}$$

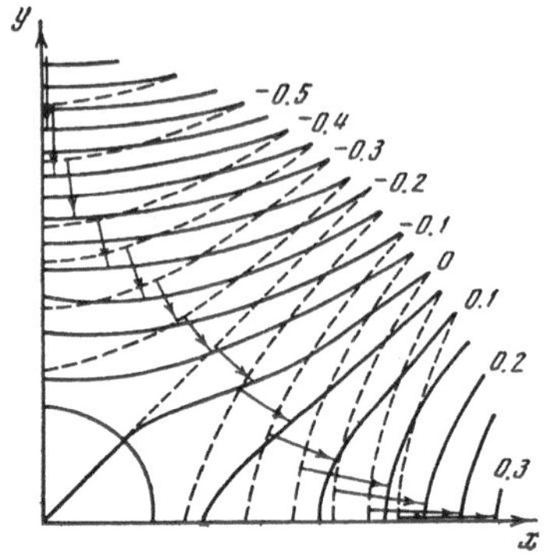

Fig. 4. Distortion of a magnetic field by a converging cylindrical wave [6]. Small circle) Wave front; solid curves) magnetic lines of force; dashed curves) initial position of these lines of force; arrow) plasma displacement. The pattern is symmetric with respect to the coordinate axes.

However, even this approximate solution reveals the most characteristic features of this type of flow. The converging wave leads to a peculiar type of accumulation for the magnetic field and the plasma current: As the null line is approached, the current j_z and the perturbation of the magnetic field increase, as is evident from Eqs. (21) and (22). The plasma motion is asymmetric: The plasma moves toward the null line along the y axis and away from it along the x axis (Fig. 4); the velocity $|v| = (v_r^2 + v_\varphi^2)^{1/2}$ increases with decreasing r [see Eqs. (23) and (24)]. In this connection the plasma density also becomes nonuniform, decreasing in two opposite quadrants and increasing in the other two [Eq. (25)]. As a result, the flow leads to a decrease in the plasma density in certain parts of the neighborhood of the null line with a simultaneous increase in the magnetic field. We note that this wave, according to the classification of MHD waves in [29, 30], is a fast magnetosonic wave.

§ 4. Formation of a Current Sheet as the Result of Two-Dimensional Plasma Flow near a Magnetic Null line

The nature of the plasma flow in the nonlinear region, $r \lesssim (\beta_1)^{1/2}$, can be determined by numerically integrating the MHD equations, (9)-(11). A qualitative examination of the nonlinear stage of the problem [3, 13, 14] on the basis of the frozen-in property of the magnetic field and the continuity of the lines of force leads to the conclusion that a current sheet of width $\sim 2(\beta_1)^{1/2}$ should develop along the x axis in the y = 0 plane, separating regions in which the magnetic fields are in opposite directions. Outside the current sheet, in a region of radius $r \sim (\beta_1)^{1/2}$, the magnetic field is nearly uniform, as it should be in the case of a plane current, and its intensity is $H \simeq h_0(\beta_1)^{1/2}$. Exact self-similar solutions [31, 15, 16] obtained for special initial and boundary conditions also permit the conclusion that a current sheet develops upon a change in a magnetic field containing a null line which is frozen in a plasma. This result was justified by Syrovatskii [17], who showed that, in the ideal case of a magnetic field frozen in a plasma, the presence of a null line in a strong magnetic field, with an electric field parallel to the null line, is a necessary and sufficient condition for the appearance of a current sheet. Syrovatskii [17] applied the name "singular null lines" to lines at which $\mathbf{H} = \nabla A = 0$ but $E = -c^{-1}\partial A/\partial t \neq 0$. The existence of singular null lines is not compatible with the MHD frozen-in condition, (10), according to which the electric field should vanish at a null line. It follows that a continuous deformation of a magnetic field and the corresponding continuous plasma motion are permitted by condition (10) everywhere except at singular null lines. Near such a line there can be no continuous flow, and a discountinuity, i.e., a current sheet, must appear in the magnetic field. Regarding the configuration of a current sheet we note that it must be situated at the singular null line in such a manner that it passes through all the secondary null lines which arise if a line current varying from zero to some finite value is placed at the original null line. The direction of this line current must be the same as that of the electric field. Accordingly, if the external field contains singular null lines, current sheets should arise near these lines in the plasma.

Syrovatskii [17] also studied the properties of a current sheet and the structure of the associated magnetic field. In seeking the field near the sheet he adopted as a boundary condition the assumption that the lines of force do not intersect the sheet and that the vector potential is constant at the surface of the sheet. This requirement follows from the continuity of the lines of force, which are conserved during their deformation because the magnetic field is frozen in the plasma. As a result, the current sheet which develops under these conditions near a singular null line is a neutral current sheet with lines of force extended along the sheet. Syrovatskii [17] assumed the current sheet to be infinitesimally thin, i.e., to be the surface y = 0, along which a current flows. The surface-current density I(x) must be distributed over the width of the sheet in a completely different manner in accordance with the boundary conditions. Since these conditions specify that the normal components of the magnetic field vanish at the surface of the sheet, $H_y^{sh} = 0$, we obviously have

$$I(x) = \frac{c}{2\pi} H_x^{sh}(x). \quad (27)$$

Away from singular null lines and current sheets the magnetic field is described by the complex potential $F(\zeta)$ [17]:

$$H_x - iH_y = i\frac{dF}{d\zeta}, \quad (28)$$

where $\zeta = x + iy$. For a current sheet developing at a first-order null line this complex potential is

$$F(\zeta) = \frac{h_0}{2}\zeta\sqrt{\zeta^2 - b^2} - \frac{2\mathcal{J}}{c}\ln\frac{\zeta + \sqrt{\zeta^2 - b^2}}{b},$$
$$\frac{dF}{d\zeta} = h_0\frac{\zeta^2 - {}^1/_2 b^2 - 2\mathcal{J}/ch_0}{\sqrt{\zeta^2 - b^2}}, \quad (29)$$

where b is the half-width of the sheet and $\mathcal{J} = \int_{-b}^{+b} I(x)\,dx$ is the total current in the sheet.

In general, currents can flow in a neutral sheet either parallel or antiparallel to the electric field E_z; return currents can appear and be maintained by the inhomogeneous motion of an ideally conducting plasma induced by the field E_z. The range of values of the return current is limited by two extremes. At one extreme, the total current $\mathcal{J}$ in the sheet can vanish; i.e., the return current can precisely cancel the forward current. In the other extreme there is no return current; in this second case we have a simple relation among the gradient h_0 of the initial magnetic field near the singular null line, the total current $\mathcal{J}$ in the sheet, and the sheet half-width b:

$$h_0 b^2 = 4\mathcal{J}/c. \quad (30)$$

Then we see from (29) that

$$H_x - iH_y = i\frac{dF}{d\zeta} = h_0\sqrt{b^2 - \zeta^2} \quad (31)$$

and in the limit $y \to 0$ we have

$$H_x = \pm h_0\sqrt{b^2 - x^2}; \quad H_y = 0 \quad \text{for} \quad |x| < b;$$
$$H_x = 0; \quad H_y = \pm h_0\sqrt{x^2 - b^2} \quad \text{for} \quad |x| > b. \quad (32)$$

The magnetic field configuration near the current sheet in the case in which there is no return current is shown in Fig. 5; it corresponds to the configuration found in self-similar

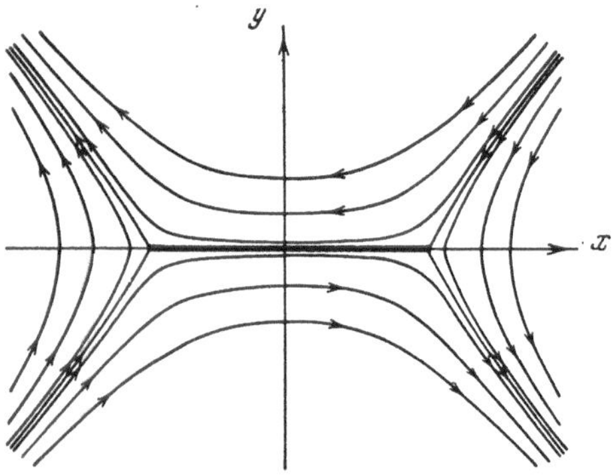

Fig. 5. Pattern of magnetic lines of force near a neutral current sheet [17] for the case in which there are no return currents.

solutions [15, 16]. From Eq. (32) and Fig. 5 we clearly see that the configuration in the case in which a neutral current sheet develops is fundamentally different from that in the case of flow involving the formation of slow-mode standing shocks [10] (Fig. 3). From (29) we can determine the components of the magnetic field near a current sheet in the other limit, i.e., in the case in which the total current through the sheet vanishes, $\mathcal{J}=0$, and for all intermediate cases. Calculating $H_x(x)$ at the surface of the sheet and using (27) we can find the distribution of the surface current density I(x) over the width of the current sheet. Figure 6 shows I(x) curves for the two limits: with a zero return current in the sheet (A) and with the return current precisely canceling the forward current (B). Curves corresponding to intermediate situations lie between curves A and B. We see from this figure that if a return current appears in the sheet it must be concentrated in the region $b/\sqrt{2} < |x| \leq b$; in the approximation of an infinitesimally thin sheet the current density becomes infinite at the ends of the sheet.

A real current sheet of course cannot be infinitesimally thin, but, strictly speaking, the theory does not give the thickness of the sheet, i.e., its dimension along the y axis. It can be assumed [32] that the sheet thickness $2a$ is equal to the distance at which the plasma pressure becomes important or at which the approximation of ideal magnetohydrodynamics breaks down, and effects due to the finite radius of curvature of the ion trajectory in the magnetic field or of the finite conductivity become important. Obviously, this dimension must be much smaller than the sheet width 2b if we are to be able to treat the current region as a sheet at all.

The formation of a neutral current sheet near a magnetic null line implies a significant increase in the magnetic energy density in this region. In fact, in a region with a diameter equal to the sheet width 2b the magnetic energy is roughly twice the initial energy of the field

Fig. 6. Distribution of the surface current density $I_z(x)$ over the width of an infinitesimally thin neutral current sheet [17]. A) The total current in the sheet is $\mathcal{J} = ch_0b^2/4$ and the return current is zero; B) the total current $\mathcal{J}$ in the sheet is zero, and the forward and return currents precisely cancel each other.

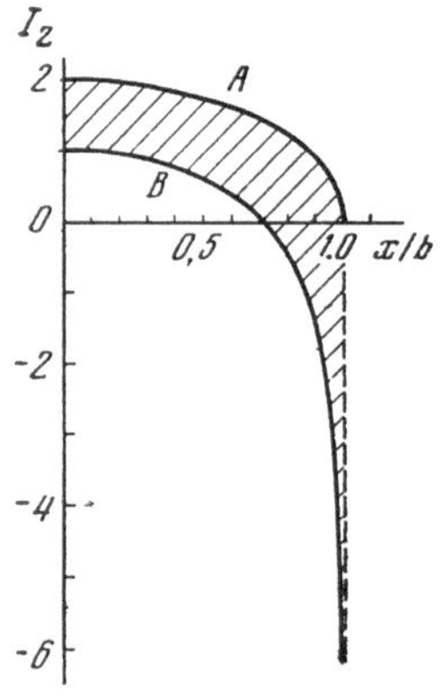

with the null line, $|\mathbf{H}| = h_0 r$, in the same region. Accordingly, the excess magnetic energy per unit length of the null line, associated with the current flowing in the sheet, is

$$\Delta W \sim \frac{1}{16} h_0^2 b^4. \tag{33}$$

The magnetic energy density near the surface of the current sheet is larger than the initial density at the same points by a factor of $(b/a)^2$. The energy concentrated in the sheet can be expended on plasma heating or, in the case of a rapid rupture of the sheet and the generation of strong pulsed electric fields, on acceleration of plasma particles.

In addition to the magnetic field configuration near a neutral current sheet there is another subject of considerable interest: the hydrodynamic plasma flow in such a region. Somov and Syrovatskii [18, 19] numerically integrated the MHD equations to find the time evolution of the plasma velocity and density distributions near a developing current sheet and near a steady-state current sheet. Their results show that low-density regions appear near the center of the sheet in both cases; the plasma density near a steady-state sheet decreases throughout the region near the sheet, i.e., over the entire length of a magnetic line of force stretching along a sheet. It thus follows from [18, 19] that a situation can arise in which the current sheet is surrounded by a plasma of an arbitrarily low density. This conclusion is of fundamental importance for the problem of particle acceleration in a plasma near magnetic null lines [20].

According to [33-36] a current sheet surrounded by a low-density plasma can become macroscopically unstable against breakup into current filaments. At the ruptures in the sheet, pulsed electric fields E_u should appear, directed along the current in the sheet. If the rupture occurs at a velocity $v \sim v_A$, the field at the rupture is

$$E_u \sim \frac{v_A}{c} H_x^{\text{sh}} \sim \frac{(H_x^{\text{sh}})^2}{\sqrt{4\pi\rho}}. \tag{34}$$

Accordingly, with a sufficiently strong magnetic field at the boundaries of the sheet, H_x^{sh}, the pulsed electric field E_u can greatly exceed the original electric field $E_z = -c^{-1}\partial A/\partial t$, which drives the plasma into motion and shapes the neutral current sheet [23]. Equation (34) shows the properties of the neutral sheet which we must try to achieve experimentally in order to obtain strong pulsed electric fields upon the rupture of the sheet.

§ 5. Effect of the Finite Plasma Conductivity on the Formation of a Neutral Current Sheet

The conclusion that a neutral current sheet forms in a plasma and that there is a significant buildup of magnetic energy near a magnetic null line is based on the concept that the magnetic field is frozen in a plasma. Any real plasma, however, has a finite conductivity, which leads to a seepage of the magnetic lines of force through the plasma or to a sort of diffusion, characterized by the magnetic viscosity coefficient

$$\nu_m = \frac{c^2}{4\pi\sigma}. \tag{35}$$

To take this process into account for the case of a finite conductivity, we must add to Eq. (10) a term describing ohmic dissipation of the magnetic field:

$$\frac{dA}{dt} \equiv \frac{\partial A}{\partial t} + (\mathbf{v}\nabla) A = \nu_m \Delta A. \tag{36}$$

We see that the magnetic field remains frozen in the plasma even when there is a finite conductivity, provided that the plasma motion with the field occurs quite rapidly, over a scale time τ much shorter than the scale time for ohmic dissipation of the magnetic field in a volume with a scale dimension l:

$$\tau \ll t_\sigma = \frac{l^2}{\nu_m}. \tag{37}$$

The quadratic dependence of t_σ on l shows that, while the frozen-in condition in (37) is easily satisfied for astrophysical objects with essentially any plasma conductivity ($l \sim 10^9$-10^{10} cm and $\tau \sim 10^2$-10^3 sec, e.g., for solar flares), in the laboratory ($l \simeq 1$-10 cm) the field remains frozen in the plasma only for an extremely short time.

This tremendous difference in spatial scales, in particular, rules out an exact simulation of astrophysical processes in the laboratory. This does not mean, however, that it is impossible to carry out an experimental study of analogous processes in the laboratory: It is possible to make use of the principle of limited modelling [37]. According to this principle, if some dimensionless parameter in the plasma in space satisfies the condition $R \gg 1$, then this parameter must also satisfy the condition $R \gg 1$ in the laboratory, but it need not be on the same order of magnitude as in space. In particular, the dimensionless parameter which we examine to determine whether the magnetic field remains frozen in the plasma is the magnetic Reynolds number

$$\mathrm{Re}_m = 4\pi\sigma l v / c^2, \tag{38}$$

where v is the scale velocity of the plasma motion. In space the condition $\mathrm{Re}_m \gg 1$ is always satisfied, and to ensure that the magnetic field is frozen in the plasma in the laboratory we must arrange the inequality $\mathrm{Re}_m \gg 1$, which is equivalent to (37)

Whether this inequality holds depends on the magnitude of the conductivity σ under the particular experimental conditions. If the plasma conductivity is governed by Coulomb collisions between particles, the conductivity is governed primarily by the temperature of the plasma electrons [38]:

$$\sigma = 1.76 \cdot 10^{13} \frac{T_e^{3/2}}{(\Lambda/10)\, Z}, \tag{39}$$

where T_e is the electron temperature in electron volts, Z is the charge of the plasma ions, and Λ is the Coulomb logarithm (usually $\Lambda \sim 10$-20).

However, if the current density exceeds a certain critical value, the so-called anomalous plasma resistance appears: Instabilities developing in the plasma cause a significant loss of the directed velocity of electrons because of the scattering of electrons in the fields of the oscillations which are excited [39, 40]; i.e., the electrons are subject to an additional friction force, and the plasma conductivity can as a result fall many orders of magnitude below the conductivity in (39), governed by Coulomb collisions. This phenomenon has been observed in several experiments, e.g., those of [41-47], and has furnished the basis for an extremely effective method of plasma heating, involving the driving of instabilities in the plasma (the so-called turbulent heating method) [48].

There are several instabilities which can lead to an anomalous resistance. In a nonisothermal plasma with $T_e > T_i$, where T_i is the ion temperature, ion-acoustic waves can be excited if the current velocity of the electrons exceeds the ion-acoustic velocity: $u > c_s = (T_e/M_i)^{1/2}$, where M_i is the ion mass. Quasilinear theory has been worked out for the ion-acoustic instability of an electric current in a plasma [49, 50]. If the drift velocity of the electrons with respect to the ions exceeds the electron thermal velocity, $u > v_{Te}$, the Buneman instability can occur [51]. The Buneman instability leads to rapid electron heating and a violation of the condition $u > v_{Te}$, and this instability ultimately converts into ion sound.

Under the condition $\omega_H < \omega_{0e}$, where $\omega_H = eH/mc$ is the electron gyrofrequency and $\omega_{0e} = (4\pi ne^2/m)^{1/2}$ is the electron plasma frequency, an electric current flowing perpendicular to the magnetic field in a plasma does not significantly alter the nature of the ion-acoustic instability, but it does lead to the appearance of new wave branches: the so-called modified Buneman instability [52, 53], the electron-acoustic instability [54], and the electron cyclotron instability (Bernstein modes [55-57]). These waves can grow in a plasma even if $T_e \leq T_i$. Finally, in an inhomogenous plasma with a current flowing across the magnetic field, an instability like ion sound but of a drift nature can arise [58].

The various types of instabilities affecting the plasma conductivity lead to general equations based on momentum exchange between electrons and waves [59]. The plasma conductivity can be written in the usual manner,

$$\sigma = \frac{ne^2}{m\nu_{eff}}, \tag{40}$$

where ν_{eff} is the effective frequency of electron collisions insofar as momentum loss is concerned; in principle, this frequency should be calculated from the conservation law for the total momentum of the system, including electrons and waves [59]. The momentum lost by electrons is ultimately transferred to ions. For the ion-acoustic and Buneman instabilities we have the following approximate equations [49, 50]:

$$\nu_{eff} \simeq \omega_{0e} \frac{\overline{W}}{nT_e}, \tag{41}$$

where $\overline{W}$ is the average energy density of the waves in the plasma and nT_e is the thermal-energy density in the plasma.

The turbulent conductivity of the plasma can be estimated from (40) if we know the relationship between ν_{eff} and the plasma properties. The following theoretical equation has been derived for ion sound [60]:

$$\nu_{eff} \simeq 10^{-2}\omega_{0e} \frac{u}{c_s} \frac{T_e}{T_i}. \tag{42}$$

Corresponding equations for ν_{eff} are also available [54, 56, 59] for the case of a current flowing across a magnetic field. The conductivity σ apparently reaches its lowest value upon excitation of the Buneman instability:

$$\sigma \simeq \frac{1}{2}\left(\frac{M}{m}\right)^{1/3} \omega_{0e}. \tag{43}$$

Experiments on the anomalous plasma resistance in a longitudinal magnetic field [46, 47, 61] have revealed the basic tendencies in the behavior of the turbulent conductivity as a function of the electric field, the density, and the mass composition of the plasma. The turbulent conductivity is several orders of magnitude below the Coulomb conductivity, (39), and increases with increasing plasma density.

A possible role of the onset of small-scale instabilities near a magnetic null line was first pointed out in [11, 12, 62].

Since the formation of a neutral current sheet is strongly related to the flow of an electric current in a plasma (the current at the wave front and then the current in the sheet), it is extremely probable that in this case instabilities will be excited which will lead to a decrease in the conductivity and will fundamentally change the nature of the plasma flow and the sheet-formation process. It is therefore necessary to carry out special experiments to determine the conditions under which current sheets can be formed in a plasma in the laboratory.

CHAPTER II

EXPERIMENTAL APPARATUS FOR STUDYING THE FORMATION OF A CURRENT SHEET NEAR A MAGNETIC NULL LINE

§ 1. Basic Requirements Which Must Be Met by the Experimental Apparatus [63]

In designing the apparatus our goal was to arrange experimental conditions providing the best approximation of the conditions under which a current sheet is formed in the theory [13-17]. Let us examine the particular requirements which must be met by the apparatus.

1. For experiments on two-dimensional plasma flow in a two-dimensional magnetic field with a null line, discussed in Chapter I, the apparatus must have a geometry such that all the conditions are homogeneous along the direction of the null line, the magnetic lines of force lie in planes perpendicular to the null line, and the plasma motion also occurs in these planes. Two-dimensional magnetic fields can evidently be produced by current-carrying straight conductors whose length is much larger than their separation. In the simplest case, two straight conductors would carry currents identical in magnitude and direction, and between these conductors there would be a first-order null line. Accordingly, the length of the apparatus along the null line (the Oz axis) must be much larger than the transverse dimension of the apparatus, and the plasma properties – the density, temperature, and profile in the (x, y) plane must remain essentially constant along the coordinate z. The electric field E_z, which causes the two-dimensional plasma flow, must also be homogeneous along z.

2. Theory for the formation of a current sheet has been worked out in the MHD approximation, in which the plasma is treated as a continuum and which is valid, strictly speaking, if the mean free path is much shorter than the scale dimension of the plasma, l, and if many collisions occur over the scale time for the process, τ [29, 64]. However, the hydrodynamic description is also valid for a low-density, collisionless plasma [65-67] if the plasma is quasineutral, i.e., if its scale dimension l is much larger than the Debye length,

$$l \gg r_d \simeq \frac{v_{T_e}}{\omega_{0e}}, \tag{44}$$

and if the gyroradii of the particles and their revolution periods in the magnetic field are small in comparison with the spatial and temporal scales, respectively:

$$l \gg r_{H_i} = \frac{v_i M_i c}{ZeH}, \tag{45}$$

$$\tau \gg \frac{2\pi}{\Omega_H} = \frac{2\pi M_i c}{ZeH}. \tag{46}$$

Here M_i and v_i are the ion mass and velocity; conditions (45) and (46) are written for ions, since these conditions are generally more stringent for ions. In an inhomogeneous magnetic field with a null line, conditions (45) and (46) are always violated in a region in which the magnetic field is weak, so it is desirable to keep such regions small in size; specifically, an effort should be made to maximize the magnetic field gradient near the null line.

3. To satisfy the MHD condition that the displacement currents be small we must ensure that the refractive index of plasma is large [64, 65]. In the case of comparatively slow processes satisfying (46) this requirement reduces to

$$N^2 = 1 + \frac{\omega_{0e}^2}{\omega_H \Omega_H} = 1 + \frac{4\pi n M_i c^2}{H^2} \gg 1. \tag{47}$$

Here N is the refractive index of the plasma, ω_{0e} is the electron plasma frequency, and ω_H and Ω_H are the electron and ion gyrofrequencies. If conditions (45) and (46) are to hold at the same time as condition (47), the plasma density n must be quite large.

4. As was mentioned in Chapter I, the formation of a current sheet was treated theoretically in the approximation of a strong magnetic field,

$$\frac{H}{\sqrt{4\pi\rho}} \gg \sqrt{\frac{\gamma p}{\rho}}. \tag{14'}$$

Since this condition is violated as $H \to 0$, the magnetic field in the apparatus must increase rapidly with distance from the null line.

5. In the case of a finite conductivity, the condition ensuring that the magnetic field is frozen in the plasma reduces to the requirement that the scale time for the process, τ, be small in comparison with the scale time for ohmic dissipation of the magnetic field, t_σ, in (37):

$$\tau < \frac{4\pi\sigma l^2}{c^2}. \tag{47'}$$

With $l \sim 1$ cm and $\sigma \sim 10^{14}$ esu the scale time for the formation of a current sheet near a null line should be less than 1 μsec.

So that we would be able to vary the experimental conditions independently over broad ranges, we provided three separate, self-contained systems for producing the magnetic field with a null line, for producing the plasma in this field, and for producing the electric field to cause the two-dimensional plasma motion.

§ 2. System for Producing a Magnetic Field with a Null Line

To optimize the system for producing this field we write the vector potential of n currents $\mathcal{J}$, identical in magnitude and direction, at the lines $r = a$, $\varphi_k = 2\pi k/n$ ($k = 0, 1, \ldots, n-1$), $n \geq 2$:

$$A_z = -\frac{\mathcal{J}}{c}\{\ln[1 - 2r_1^n \cos n\varphi + r_1^{2n}] - C\}, \tag{48}$$

where $r_1 = r/a$ and where C is a constant governed by the complete current configuration, including the completion of the current path at large distances. The line $r_1 = 0$ is a magnetic null line of order $n-1$ ($H = 0$ at $r_1 = 0$). Near the null line, i.e., at $r_1^n \ll 1$, the vector potential and the magnetic field are

$$A_z = \frac{2\mathcal{J}}{c}\left[\left(\frac{r}{a}\right)^n \cos n\varphi + \frac{C}{2}\right],$$
$$\mathbf{H} = [\nabla A_z \cdot \mathbf{e}_z] = \left\{\frac{1}{r}\frac{\partial A_z}{\partial \varphi};\ -\frac{\partial A_z}{\partial r};\ 0\right\} = -\frac{2\mathcal{J} n}{ca}\left(\frac{r}{a}\right)^{n-1}\{\sin n\varphi;\ \cos n\varphi;\ 0\}. \tag{49}$$

We see from (49) that the magnetic field near a first-order null line ($n = 2$) increases most rapidly – linearly. For $(r/a)^2 \ll 1$ we find equations equivalent to (5) and (6):

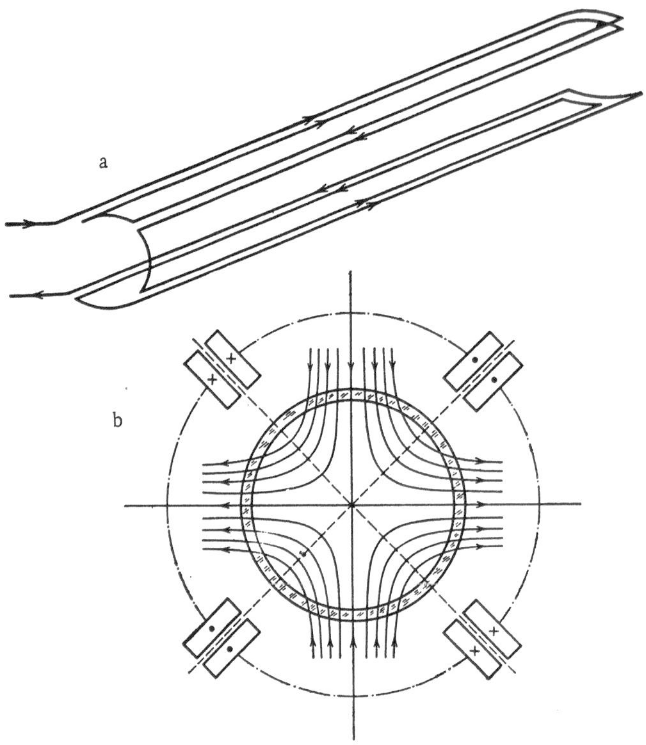

Fig. 7. System of conductors producing a magnetic field with a null line (a straight quadrupole). a) Current flow; b) vacuum chamber in the quadrupole magnetic field, i.e., cross section perpendicular to the null line. The rectangles are the conductors; the solid curves show the magnetic lines of force.

$$A_z = \frac{h}{2} r^2 \cos 2\varphi + C_1,$$
$$\mathbf{H} = -hr\{\sin\varphi;\ \cos\varphi;\ 0\}, \tag{50}$$

where the magnetic field gradient h is governed by the magnitude of the currents $\mathcal{J}$ and by the distance between them, specifically, $h = 4\mathcal{J}/ca^2$.

A first-order null line with a gradient twice as large for the same values of $\mathcal{J}$ and a can be obtained by using currents flowing in both the "forward" and "reverse" direction, at the lines ($r = a$, $\varphi = 0; \pi$) for $+\mathcal{J}$ and at the lines ($r = a$, $\varphi = \pi/2; 3\pi/2$) for $-\mathcal{J}$. In other words, we make use of a system of conductors corresponding to a straight quadrupole, for which the vector potential is

$$A = -\frac{\mathcal{J}}{c}\{\ln[1 - 2r_1^2\cos 2\varphi + r_1^4] - \ln[1 + 2r_1^2\cos 2\varphi + r_1^4]\}; \tag{51}$$

in the case $r/a \ll 1$ this vector potential reduces to (50), where h is understood to be twice its value,

$$h_0 = \frac{8\mathcal{J}}{ca^2}. \tag{52}$$

Accordingly, to provide the fastest increase in the magnetic field near a null line we should choose a straight quadrupole to produce the magnetic field.

We see from Eq. (52) that there are two ways to increase the magnetic field gradient: by raising the current in the quadrupole conductors or by reducing the transverse dimension a of the quadrupole. This latter measure seems particularly attractive, since in this case it is in principle possible to produce a high gradient without increasing the current supply. However, if the transverse dimension is reduced, the frozen-in condition for the magnetic field may be violated; furthermore, it may become physically difficult to carry out measurements in the plasma. As a result we adopted the following geometry: The quadrupole conductors are 76 cm long, arranged in a cylinder 11.2 cm in diameter (a = 5.6 cm). The current in the conductors is produced by a bank of 20 type IM 5/150 capacitors with a total capacitance of $3 \cdot 10^3$ μF; the stored energy is W = 37.5 kJ. The bank, charged to $U \leq 5$ kV, is discharged by means of a controllable vacuum switch into the conductors producing the quadrupole field. Structurally, each of the four arms of the quadrupole (Fig. 7) consists of two conductors (flat copper busbars 20 × 6 mm in cross section), separated by 1 mm of textolite insulation over their entire length. All eight of these conductors are connected in series (Fig. 7a), so that the inductance of the quadrupole is increased [68], and the relative contribution of the inductances of the capacitor bank, the connecting cables, and the switch is reduced; i.e., the relative importance of the magnetic fringing fields is reduced. On the other hand, this increase in the inductance makes it possible to increase the duration of the quadrupole field (the half-period is T/2 = 400 μsec). The other processes – the plasma formation and motion – are characterized by much shorter scale times and occur at the peak of the quadrupole field, which is thus a quasisteady-state field in comparison. The gradient of the magnetic field in the system reaches $h_0 = 5 \cdot 10^3$ Oe/cm.

§ 3. Plasma Injection into a Magnetic Field with a Null Line [69]

Multipole magnetic fields are widely used in experiments on controlled thermonuclear fustion (see, e.g., [70-74]). The basic features of these fields – the increase of the field toward the periphery and the negative curvature of the lines of force – make these fields extremely convenient for isolating a plasma from the chamber walls and for confining it near the axis, where the field is weak. In designing the present experiments, on the other hand, we were faced with precisely the opposite problem: that of arranging the uniform filling of the plasma region by an inhomogeneous magnetic field with a null line. Accordingly, the very properties which make a system a good magnetic trap would complicate our task.

A cylindrical glass vacuum chamber, with inside diameter of 6 cm, is placed in a quadrupole magnetic field in such a manner that the chamber coincides with the magnetic null line (Fig. 7b). The chamber is evacuated to $2 \cdot 10^{-6}$ torr. In a first series of experiments the plasma was produced by an arc-type injector with internal ignition [75], in which a discharge occurred along a plastic surface between two molybdenum electrodes. Here the source had a capacitance C = 0.2 μF and was charged to a voltage $U \leq 6$ kV.

If a plasma injector lies outside a multipole magnetic field and injects plasma into this field along a null line, the field is known to operate as a limiter, selecting from the plasma stream a narrow jet of radius satisfying

$$\frac{\rho v^2}{2} \geqslant \frac{H^2}{8\pi} = \frac{(h_0 r)^2}{8\pi}, \tag{53}$$

where ρ is the density and v is the velocity of the plasma stream [74]. Accordingly, in order to produce as broad as possible a plasma distribution in the plane perpendicular to the null line, the injector was placed in a region at the end of the chamber where the quadrupole magnetic field was longitudinally homogeneous. The plasma stream propagated in the direction of the null line, filling the entire length of the chamber in 10 μsec. The plasma density distribution at various points along the length and cross section of the chamber was measured by a shielded electric probe, operating in the saturation ion current regime. The probe was inserted into the chamber from the end opposite the injector end, so that the entrance aperture of the probe, $S = 0.8\ mm^2$ in area, faced the incoming plasma stream. A special positioning device made it possible to place the probe anywhere within the vacuum chamber. This probe measured the plasma flux (n**v**), and since the plasma-production time for this particular source was short [75], we were able to determine the plasma velocity **v** and thus the plasma density n from the transit time of the plasma from the source to the probe.

The measurements revealed that the plasma velocity was essentially constant over the chamber cross section, i.e., that the stream moved as a whole, even though the motion was across a magnetic field which was very inhomogeneous over the cross section (the field was H = 0 at the axis, while at the chamber wall it reached ~15 kOe at the maximum gradient). Figure 8 shows the distribution of the plasma density over the cross section at a point l = 58 cm from the source at the time corresponding to the arrival of the maximum plasma density. The plasma boundaries essentially coincide with lines of force of the quadrupole field; the density is maximal at the null line. Along a line connecting conductors of the quadrupole in which the current was flowing in the same direction (AA' in Fig. 8) the plasma density varied slowly in a region 2-3 cm in size near the null line, thereafter decaying rapidly, by two orders of magnitude at a distance ~6 mm (Fig. 9). As the field gradient was increased the distribution be-

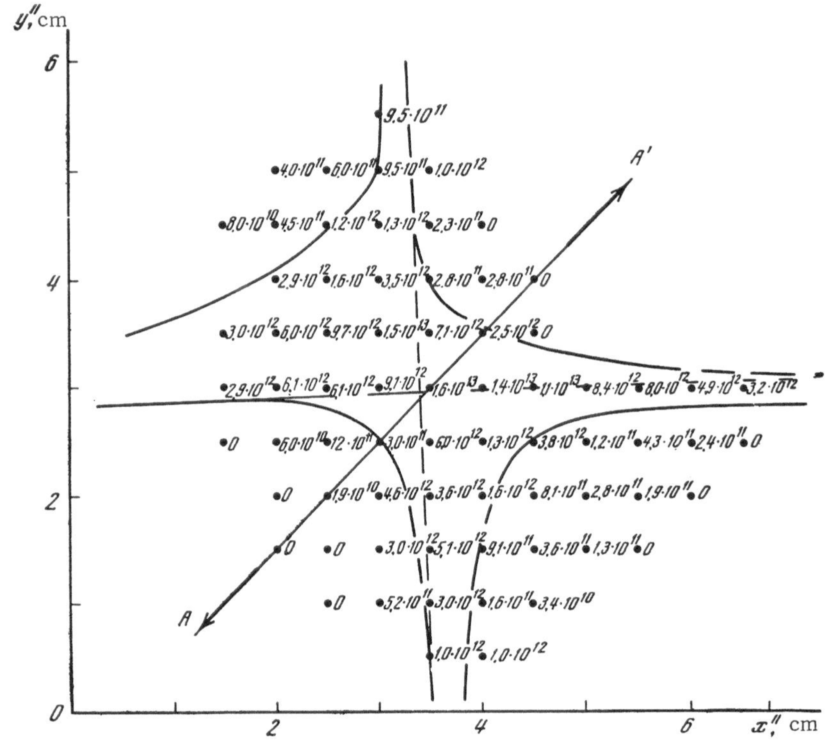

Fig. 8. Profile of the plasma density produced by a plane injector in the transverse cross section of the vacuum chamber in a quadrupole magnetic field with a gradient $h_0 = 10^3$ Oe/cm at a point l = 58 cm from the plasma source.

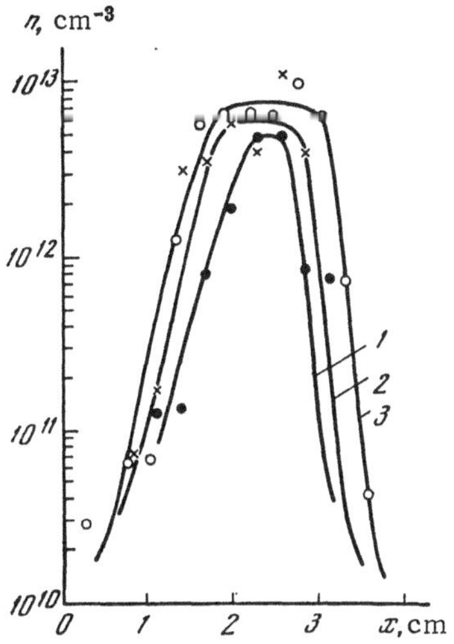

Fig. 9. Profile of the plasma density along line AA' (Fig. 8) at a point l = 58 cm from the plasma source for various magnetic field gradients h, kOe/cm: 1) 2.0; 2) 1.0; 3) 0.5.

came narrower, with the plasma concentrating near the null line; in this case the density was slightly lower (Fig. 9). These changes in the density profile were apparently due to the nature of the injector operation in the quadrupole magnetic field. Near the magnetic field "apertures" (between the conductors of the quadrupole) the plasma density was nonvanishing all the way out to the chamber walls (Fig. 8). The nature of the plasma motion in a magnetic field with a null line was found to agree well with the results in [76], even though the field gradient in the present experiments was an order of magnitude larger.

The properties of the plasma stream in the quadrupole stream can be explained by taking into account the electric field which arises because of the plasma polarization. At the instant of plasma production in the injector the magnetic field is clearly displaced, and the plasma can move a certain distance away from the null line in the transverse direction, simultaneously acquiring a velocity $\mathbf{v}_{z0}$ along the chamber axis because of electrodynamic forces. However, as soon as the magnetic field penetrates into the plasma, the electrons and ions begin to be deflected in opposite directions, so that surface charges appear at the plasma boundaries, and a polarization electric field E_p arises. If the dielectric constant of the plasma is large, i.e., if

$$\varepsilon = 1 + \frac{4\pi n M_i c^2}{H^2} \gg 1, \tag{47'}$$

then the surface charge density increases until we reach a situation

$$\mathbf{E}_p = -\frac{1}{c}[\mathbf{v}_{z_0} \times \mathbf{H}], \tag{54}$$

after which the plasma drifts along the axis at $\sim \mathbf{v}_{z0}$. If condition (54) is violated because some factor suppresses the polarization and reduces the field E_p, the particles are again deflected by the magnetic field, so that the surface charge density increases. Such a mechanism leading to the regeneration of the polarization field can evidently operate until condition (47') is satisfied. If this condition is violated because of a density decrease, plasma motion across the magnetic field is prevented [77]. As the plasma moves along the null line of the quadrupole field $\mathbf{H} = [\nabla A \cdot \mathbf{e}_z]$ at an initial velocity $\mathbf{v} = V \cdot \mathbf{e}_z$ condition (54) can be written

$$\mathbf{E}_p = -\nabla\varphi = \frac{1}{c}[\mathbf{v} \times \mathbf{H}] = -\frac{V}{c}\nabla A, \tag{55}$$

from which we find

$$\varphi = \frac{V}{c} A + \varphi_0. \tag{56}$$

In other words, the magnetic lines of force (A = const) become equipotentials of the polarization electric field as the plasma moves along the null line.

Surface charges can flow along the magnetic lines of force, producing a polarization field where the plasma density was originally low. This mechanism was apparently responsible for the plasma penetration into the magnetic "apertures" and the subsequent escape of plasma to the chamber walls.

When the glass vacuum chamber was replaced by a metal chamber, the suppression of the polarization prevented the plasma from moving in this manner, and the plasma rapidly escaped to the walls. This result confirmed that it was in fact the polarization of the plasma which was responsible for the plasma motion across the strong, inhomogeneous quadrupole field.

The longitudinal plasma loss was slight, so that the density varied only slightly in the longitudinal direction. In a quadrupole with a gradient of 1 kOe/cm, for example, the maximum plasma density at the null line 30 cm from the injector was $\simeq 1.9 \cdot 10^{13}$ cm^{-3}, which is only 1.2-1.3 times as large as that at 60 cm; the plasma profile in the transverse cross section was like that shown in Fig. 8. It should be noted that, in the absence of the quadrupole field, the plasma density 30 cm from the injector was homogeneous over the chamber cross section, with a maximum value of $\sim 1.5 \cdot 10^{12}$ cm^{-3}; at 60 cm, the density fell by a factor of three to $\sim 5 \cdot 10^{11}$ cm^{-3}.

In the course of the experiments the plasma density was adjusted by adjusting the capacitance of the injector or the charging voltage. The electron temperature in the plasma produced by the injector was $T_e \sim 5$ eV [78]; i.e., the plasma conductivity due to Coulomb collisions reached $\sigma_h \sim 10^{14}$ esu. The maximum frequency of binary electron–ion collisions, corresponding to the maximum density $n \sim 10^{13}$ cm^{-3} near the null line, was $\nu_h \sim 2 \cdot 10^7$ sec^{-1}. The condition of a strong magnetic field or of a small gaskinetic pressure, (14'), was satisfied everywhere except in a small region less than 1 mm in diameter in size near the null line, in which the quadrupole field gradient was $h_0 \gtrsim 500$ Oe/cm. The refractive index of the plasma for low-frequency perturbations, (47), was much larger than unity at all magnetic field gradients up to the maximum gradient of $h_0 = 5 \cdot 10^3$ Oe/cm, provided that the plasma density exceeded $n > 10^{10}$ cm^{-3}. Accordingly, the requirements stated in §1 of the present chapter were met by the properties of this plasma.

§ 4. System for Producing the Induction Electric Field E_z along the Magnetic Null Line

After the plasma filled the entire chamber, the induction electric field E_z, parallel to the null line, was turned on. This field was produced in an oscillatory discharge of a capaci-

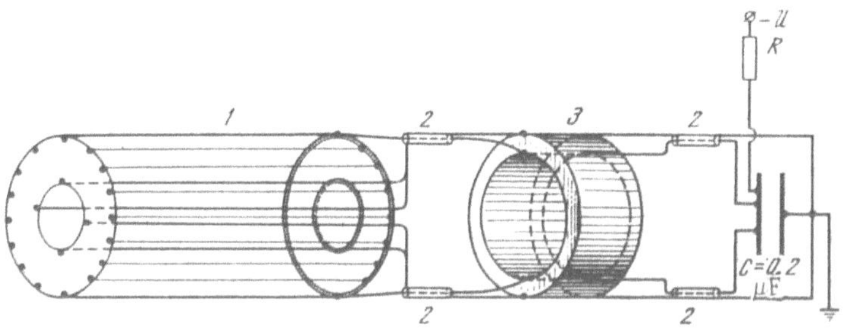

Fig. 10. System producing the induction electric field E_z. 1) System of conductors producing E_z; 2) coaxial cables; 3) vacuum switch.

tance of $C = 0.2\ \mu F$ ($U_{cha} \leq 50$ kV) into a system of straight conductors 50 cm long, arranged along two coaxial cylindrical surfaces with radii $r = a = 5$ cm (four conductors) and $r = b = 15$ cm (16 conductors) (Fig. 10). The axis of these cylinders coincided with the magnetic null line and the axis of the vacuum chamber. The currents in all the inner conductors flowed in the same direction; the currents in all the outer conductors flowed in the opposite direction. The inductance of this system of conductors was ~50% of the total inductance of the circuit, 230 cm; the resonant frequency of the circuit was $f_0 \simeq 0.75$ MHz. With this configuration it was possible to produce an electric field homogeneous over the length and cross section of the vacuum chamber; the fringing magnetic field of this system of conductors within the vacuum chamber, i.e., at $r \leq 3$ cm, was much lower than the quadrupole field.

Using Eq. (48) and neglecting edge effects, we can write the vector potential of this system of currents as

$$A_1 = -\frac{\mathcal{J}_1}{4c}\left[\ln\left(1 - 2r_1^4 \cos 4\varphi + r_1^8\right)\right] + \frac{\mathcal{J}_1}{16c}\left[\ln\left(1 - 2r_2^{16} \cos 16\varphi + r_2^{32}\right)\right] + \frac{2\mathcal{J}_1}{c}\ln\frac{b}{a}, \tag{57}$$

where $\mathcal{J}_1$ is the total current flowing through all the conductors, $r_1 = r/a$, and $r_2 = r/b$. Within the vacuum chamber, i.e., at $r \leq 3$ cm, the first two terms in (57) are small in comparison with the last term. For example, at $r = 3$ cm and with $4\varphi = 1$, i.e., where the contribution of the first term in Eq. (57) is maximal, this term is only ~0.03 of the last term. It is even more legitimate to neglect the contribution of the second term at $r \leq 3$ cm, so that within an error of ~3% we can assume the vector potential of this system of conductors to be

$$A_1 = \frac{2\mathcal{J}_1}{c}\ln\frac{b}{a} \tag{58}$$

within the vacuum chamber, essentially independent of the spatial coordinates. However, in this case the magnetic field $\mathbf{H}_1 = [\nabla A_1 \times \mathbf{e}_z]$ is weak, and the electric field is homogeneous:

$$E_z = -\frac{1}{c}\frac{\partial A_1}{\partial t} = -\frac{2}{c^2}\ln\frac{b}{a}\frac{d\mathcal{J}_1}{dt}. \tag{59}$$

In a typical experiment, with $\mathcal{J}_1 = 30$ kA, the magnetic field of this system of conductors at the chamber wall ($r = 3$ cm) was $H_1 = 260$ Oe; at $r = 2$ cm it was $H_1 = 77$ Oe. These values of H are much smaller than the quadrupole field at the same points, even at the maximum gradients ($h_0 = 500$ Oe/cm). At a current $\mathcal{J}_1 = 30$ kA the induction electric field in the chamber was $E_z \simeq 300$ V/cm. Figure 11 shows the vector potential A_1 as a function of the distance to the null line, r, over the integral $0 \leq r \leq b$, according to Eq. (57), for three values of the angle φ. Since, according to (4), the difference between the values of A_1 at two lines of force is equal to the magnetic flux between them, these curves clearly show that less than 3% of the

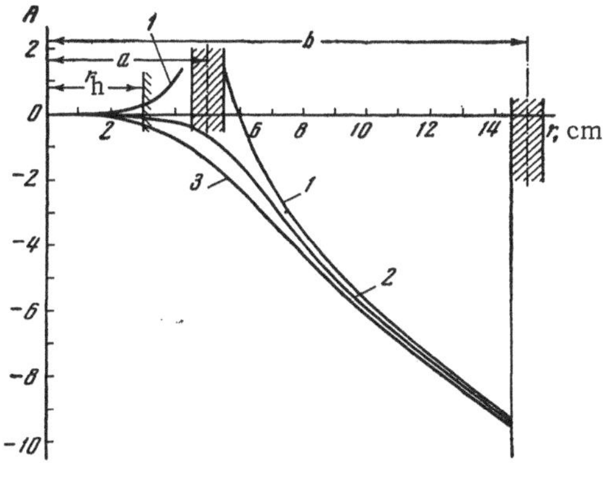

Fig. 11. Vector potential A_1 of the system producing the induction electric field E_z as a function of the distance to the null line, r. Here $r_h = 3$ cm is the inside radius of the vacuum chamber, $a = 5$ cm is the radial position of the four conductors carrying currents flowing in one direction, and $b = 15$ cm is the radial position of the 16 conductors carrying currents in the opposite direction. 1) $\varphi = 0$; 2) $\varphi = \pi/8$; 3) $\varphi = \pi/4$.

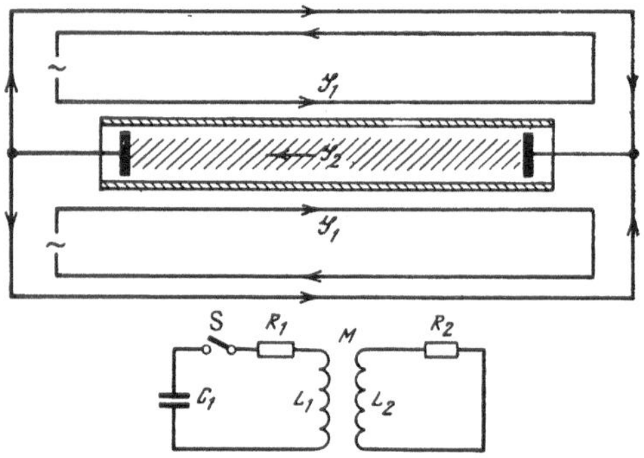

Fig. 12. System for turning on the induction electric field; equivalent circuit of the two coupled circuits.

total magnetic flux of this system of conductors penetrated into the vacuum chamber, $r \leq r_h$. This flux was concentrated primarily between the inner and outer conductors, i.e., at $a \leq r \leq b$.

The system of conductors producing the electric field in the vacuum chamber is the primary winding of an air-core transformer, and the plasma is part of the secondary winding of the same transformer (Fig. 12). Two electrodes are inserted into the vacuum chamber from its ends; they are connected by eight straight rods arranged uniformly in a cylinder of radius R = 17 cm, i.e., external to the primary circuit. The circuits are coupled by the magnetic flux linking both of them, and since there is essentially no magnetic flux within the vacuum chamber the mutual induction coefficient M should be independent of the current profile within the chamber. From the Kirchhoff equations we find the following relation between the currents for two inductively coupled circuits, with the equivalent circuits shown in Fig. 12, for a sinusoidal current $\mathcal{J}_1(t)=\mathcal{J}_{10}\exp(i\omega t)$ in the primary circuit [79]:

$$\mathcal{J}_2=-\frac{i\mathcal{J}_1\omega M}{R_2+i\omega L_2}=\frac{U_2}{R_2+i\omega L_2}, \tag{60}$$

where $U_2=-i\omega M\mathcal{J}_1$ is the voltage induced in the circuit containing the plasma when the current $\mathcal{J}_1$ changes.

The phase shift between the voltage U_2 and the current $\mathcal{J}_2$ in the secondary circuit is

$$\varphi=\arctan(\omega L_2/R_2). \tag{61}$$

We obviously have $\varphi \to 0$ if the resistance of this secondary circuit is high, $R_2 \gg \omega L_2$; then the phase shift between currents $\mathcal{J}_1$ and $\mathcal{J}_2$ is $\psi \to \pi/2$. If $R_2 \ll \omega L_2$, then $\varphi \simeq \pi/2$ and $\psi \simeq \pi$.

In turn the current $\mathcal{J}_2$ must affect the primary circuit. The results are changes in the effective inductance and the resistance of the primary circuit:

$$\begin{aligned} L_{\text{eff}} &= L_1-\frac{\omega^2M^2}{R_2^2+(\omega L_2)^2}L_2,\\ R_{\text{eff}} &= R_1+\frac{\omega^2M^2}{R_2^2+(\omega L_2)^2}R_2. \end{aligned} \tag{62}$$

Consequently, there are changes in the resonant frequency of the primary circuit and in the current $\mathcal{J}_1$. In a first approximation, neglecting damping, we write

$$\omega^2=\frac{1}{CL_{\text{eff}}};\qquad \mathcal{J}_1=U_1\sqrt{\frac{C}{L_{\text{eff}}}}. \tag{63}$$

To determine the mutual induction coefficient M, the voltage in the secondary circuit U_2, and the induction L_2 we carried out model measurements in which we simulated the plasma by metal rods with a length equal to the distance between the electrodes and with several cross-sectional shapes. In these model measurements the condition $R_2 \ll \omega L_2$ was satisfied, so that we can write

$$\mathcal{J}_2 = -\mathcal{J}_1 M/L_2, \quad L_{\mathrm{eff}} = L_1 - M^2/L_2. \tag{04}$$

By measuring the ratio $\mathcal{J}_1/\mathcal{J}_2$ and the resonant frequency of the primary circuit, we can determine M and L_2 from (64) for the various rods simulating the plasma.

As we mentioned above, the mutual induction coefficient should be independent of the shape of the rod; in fact, the measurements show that this coefficient is the same, M = 92 ± 15 cm, for all the rods used. Knowing M and L_{eff} we can immediately determine the part of the total voltage which is transferred from the primary circuit to the secondary circuit:

$$\frac{U_2}{U_1} = \frac{M}{L_{\mathrm{eff}}} \simeq 0.42. \tag{65}$$

CHAPTER III

CURRENT PROFILE NEAR A MAGNETIC NULL LINE; TURBULENT PLASMA RESISTANCE [69, 80]

§ 1. Measurements of the Total Current and the Resistance of a Plasma in a Magnetic Field with a Null Line

When the electric field E_z is applied, a current appears along the null line in the plasma in the quadrupole magnetic field. To find the magnitude and profile of this current over the first quarter-period of the field E_z, before the field changed sign, we measured the total plasma current $\mathcal{J}_2$ with a Rogowski loop around the outside of the vacuum chamber. This loop operates in the current-transformer mode [81, 82]; i.e., its signal, proportional to $\partial\mathcal{J}_2/\partial t$, is integrated by an LR circuit, where L is the inductance of the loop and R = 1.51 Ω is the resistance from which the voltage proportional to $\mathcal{J}_2$ is taken. The time constant of the integrating circuit is $2.65 \cdot 10^{-6}$ sec, the sensitivity of the loop is 26 V/kA, and the signal from the loop is fed to the plates of an OK-21 dual-trace oscilloscope. Figure 13 shows a typical oscillogram of the plasma current $\mathcal{J}_2$, recorded along with the voltage U_2 applied to the plasma, measured by a capacitive divider. We see from this oscillogram that there is essentially no phase shift

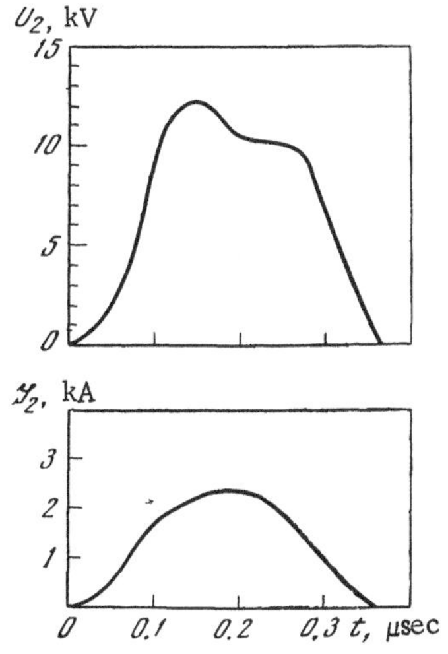

Fig. 13. Oscillograms of the voltage U_2 applied to the plasma and of the plasma current $\mathcal{J}_2$. The gradient of the quadrupole magnetic field is $h_0 = 10^3$ Oe/cm; the plasma is produced by a plane injector; $n \lesssim 10^{13}$ cm^{-3}.

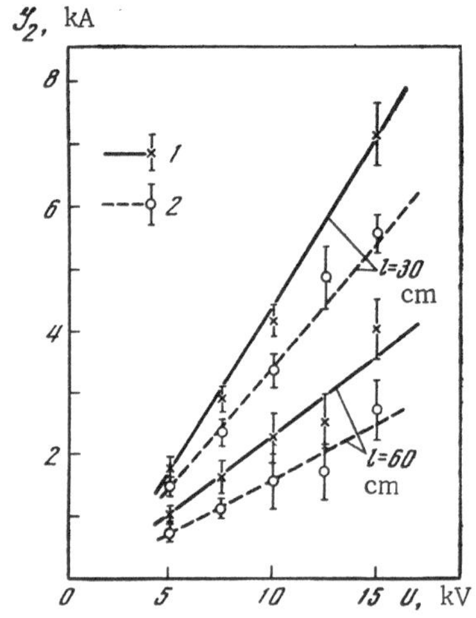

Fig. 14. Plasma current $\mathcal{J}_2$ as a function of the applied voltage for various gradients h and for various electrode separations l. The plasma is produced by a plane injector; $n \lesssim 10^{13}$ cm^{-3}. 1) $h = 10^3$ Oe/cm; 2) $h = 2 \cdot 10^3$ Oe/cm.

between the voltage and the current ($\varphi \simeq 0$); i.e., according to (61), the resistance of the circuit with the plasma is high in comparison with the inductance, $R_2 \gg \omega L_2$. The value of R_2 at the instant corresponding to the maximum current and to $\partial \mathcal{J}_2/\partial t = 0$ was determined from oscillograms of this type: $R_2 = U_2/\mathcal{J}_2$.

Measurements of the total plasma current under various experimental conditions revealed that the current decreases with increasing magnetic field gradient h and with decreasing plasma density n. For constant values of h and n the plasma current increases linearly with increasing U_2; i.e., the plasma resistance is independent of U_2, within the experimental error. Figure 14 shows the maximum current $\mathcal{J}_2$ as a function of the voltage U_2 for two values of the gradient h. To determine what was responsible for the plasma resistance – phenomenon occurring in the plasma column itself or electrode processes – we recorded dependences for two lengths of the plasma gap: 60 and 30 cm. We see from Fig. 14 that for all values of U_2 a halving of this length led to an increase of the current $\mathcal{J}_2$ by a factor of ~ 2; i.e., the plasma resistance fell by a factor of approximately two. The values of the plasma resistance R_2 is determined from the slope of the experimental $\mathcal{J}_2(U_2)$ dependences as follows: with $h = 10^3$ Oe/cm, $R_2 = 4.3$ Ω for $l = 60$ cm and 2.1 Ω for $l = 30$ cm; with $h = 2 \cdot 10^3$ Oe/cm, $R_2 = 6.0$ Ω for $l = 60$ cm and $R_2 = 2.7$ Ω for $l = 30$ cm. These results show that the resistance of the plasma circuit is not governed by electrode processes and that its distribution over the length of the plasma column is essentially uniform.

These results and those shown in Fig. 14 correspond to the maximum plasma density which could be produced in a quadrupole field with this particular injector; near the null line, at a point 60 cm from the injector, this density was $n_0 \simeq 1.7 \cdot 10^{13}$ cm^{-3}. When the density was reduced, the plasma resistance increased and the current $\mathcal{J}_2$ decreased. For example, with a magnetic field gradient of h = 950 Oe/cm and a plasma density of $n_0 \simeq 7 \cdot 10^{12}$ cm^{-3} near the null line the plasma resistance was 8.5 ± 2.5 Ω. The high resistance of the plasma gap could have been caused by two factors: Either the cross-sectional area of the current-carrying region in the quadrupole field was ~ 0.1 cm^2 with an initial plasma conductivity $\sigma_h \sim 10^{14}$ esu, or the plasma conductivity was much lower than σ_h. By measuring the profile of the current density over the plasma cross section we would be able to choose between these alternatives.

The measured total current $\mathcal{J}_2$ could be used to calculate the energy absorbed in the plasma. Since the entire voltage drop in the secondary circuit is across the plasma (since $R_2 \gg \omega L_2$), the energy absorbed by the time the current $\mathcal{J}_2$ reaches its maximum, $t_p \simeq 0.2$ μsec, is

$$W = \int_0^{t_p} U_2(t)\, \mathcal{J}_2(t)\, dt \simeq 2 \cdot 10^7 \text{ ergs}. \qquad (66)$$

If we assume that all this energy is expended on plasma heating, that the energy loss is small over the heating time, and that the total number of particles corresponds to the profile in Fig. 8, then we conclude that the plasma temperature should be $(T_e + T_i) \sim 5$ keV. Actually, of course, the plasma temperature must be much lower, primarily (it would seem) because of heat conduction along the magnetic lines of force.

§ 2. Measurements of the Current Profile in a Plasma in a Magnetic Field with a Null Line

The profile of the current density j_z flowing along the null line in the plasma was determined in the plane perpendicular to the null line by means of magnetic probes. Each of these magnetic probes is a ten-turn, single-layer coil in an electrostatic shield; the diameter and height of the probe, in its shield, do not exceed 2 mm. The signal from the probe, which is integrated by an RC circuit with a time constant of 2 μsec and which is proportional to the local value of the magnetic field, is fed to the amplifier of an OK-17M oscilloscope. These magnetic probes do not detect the quasiconstant quadrupole field; they measure only the magnetic field of the current flowing in the plasma. They are moved within a glass tube having an outside diameter of 2.8 mm, sealed into the vacuum chamber roughly halfway between the electrodes. The tube axis lies in the (x, y) plane, perpendicular to the magnetic null line, and intersects the null line at an angle of 15° with respect to the x or y axis (Figs. 15 and 16). Here the y axis is the direction along which the plasma should contract toward the null line in the quadrupole field, and the x axis is the direction along which the plasma should spread out (Chapter I). Whether the tube with the probe was moved along the x axis or the y axis depended on the direction of the electric field E_z and thus on the plasma current, on the one hand, and on the direction of the currents in the conductors producing the quadrupole field, on the other.

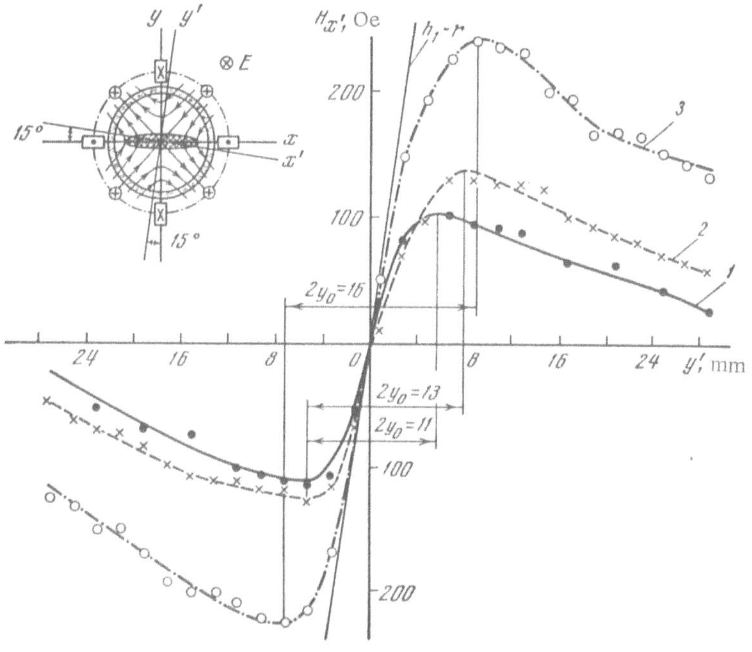

Fig. 15. Magnetic field of the plasma current, $H_{x'}(y')$. The plasma is produced by a plane injector; $n \lessapprox 10^{13}$ cm^{-3}. 1) $h_2 = 1420$ Oe/cm, E = 110 V/cm; 2) $h_1 = 640$ Oe/cm, E = 110 V/cm; 3) $h_1 = 640$ Oe/cm, E = 220 V/cm.

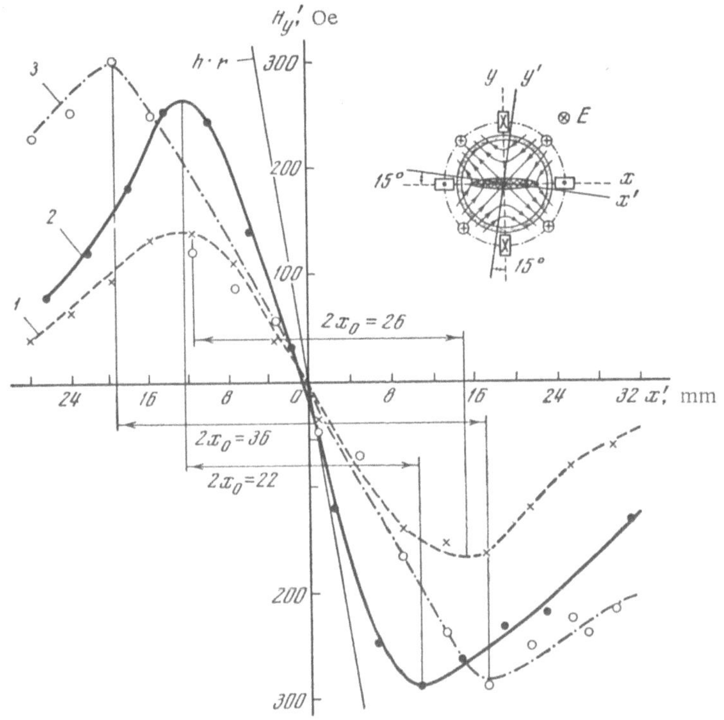

Fig. 16. Magnetic field of the plasma current, $H_{y'}(x')$. The plasma is produced by a plane injector, $n \leqslant 10^{13}\ cm^{-3}$. 1) $h_1 = 640$ Oe/cm, E = 110 V/cm; 2) $h_2 = 1420$ Oe/cm, E = 220 V/cm; 3) $h_1 = 640$ Oe/cm, E = 220 V/cm.

Electrodynamic force should cause the plasma to spread out along the line connecting the quadrupole-field conductors carrying currents in the same direction as the plasma current (along the Ox axis) and should cause contraction of the plasma in the perpendicular direction. A reversal of the current in the system of quadrupole-field conductors or a reversal of the electric field led to a rotation of the entire current distribution through $\pi/2$ in the (x, y) plane. It is convenient to introduce a coordinate system x', y', z rotated ~15° around the Oz axis with respect to the x, y, z system; then the lines along which the magnetic probes are moved coincide with the x' and y' axes. In the experiments, probes in two orientations were used, to measure the components of the magnetic field in the (x, y) plane respectively parallel to the axis of the tube containing the magnetic probe, $H_{x'}(x')$ and $H_{y'}(y')$, and perpendicular to this axis, $H_{x'}(y')$ and $H_{y'}(x')$.

TABLE 1

h, Oe/cm	E_z, V/cm	J_z, kA	$2x_0$, cm	$2y_0$, cm	σ, esu	$\left(\frac{\partial H_{x'}}{\partial y'}\right)_0$, Oe/cm	$-\left(\frac{\partial H_{y'}}{\partial x'}\right)_0$, Oe/cm	$(\text{rot } H)_0$, Oe/cm	j_{z0}, A/cm²	σ_0, esu	$r_c = \left(\frac{2M_i c^2}{e}\frac{E}{h^2}\right)^{1/3}$, cm	
											H+	C+
640	220	3.0	3.6	1.6	$2.5 \cdot 10^{12}$	500	180	680	540	$2.5 \cdot 10^{12}$	4,9	11,5
640	110	1.5	2.6	1.3	$4 \cdot 10^{12}$	320	155	475	380	$3.5 \cdot 10^{12}$	3.9	9.1
1420	220	1.8	2.2	1.2	$3 \cdot 10^{12}$	660	330	990	790	$3.6 \cdot 10^{12}$	2.9	6.7
1420	110	0.9	2.0	1.1	$3.7 \cdot 10^{12}$	330	216	546	430	$3.9 \cdot 10^{12}$	2.3	5.3

It has been shown theoretically [6, 14] (see also Chapter I) that the electric field which appears at the boundaries of the plasma in a quadrupole magnetic field cannot instantaneously penetrate into the plasma. The electric field E_z must be carried by a magnetosonic wave, and an electric current can appear at a point in the interior of the plasma only after the wave has reached this point. Under the conditions of these experiments, however, the time $\tau_A \simeq (4\pi\rho/h_0)^{1/2}$ required for the wave to reach the null line did not exceed ~30 nsec, i.e., was short in comparison with the time scales under study. Accordingly, the temporal changes in the magnetic field at different spatial points were in phase with the total current $\mathcal{J}_z$ flowing in the plasma. The spatial profiles of the magnetic field shown in Figs. 15 and 16 refer to the maximum of the plasma current $\mathcal{J}_z$ during its first half-period. Direct measurements showed that the current distribution in the plasma was independent of the coordinate z.

We see from the profiles of the magnetic field of the plasma current (Figs. 15 and 16) that the current flows primarily through the center of the chamber, near the null line, and that the current distribution is not cylindrically symmetric. If we define the size of this current region along the x' or y' axis as the distance between the positive and the negative maxima of the $H_{x'}(y')$ or $H_{y'}(x')$ curve, we see immediately by comparing the curves in Figs. 15 and 16 corresponding to identical conditions that the current is compressed along the Oy axis and stretched out along the Ox axis. In other words, the current in the quadrupole magnetic field has assumed the form of a sheet. This result also follows from measurements of the longitudinal component of the magnetic field, $H_{x'}(x')$. The point at which the magnetic-field profiles in Figs. 15 and 16 change sign is the intersection of the line x' = 0 or y' = 0 with the symmetry plane of the current sheet. This plane coincides very accurately with the magnetic null line, since the resulting curves were very symmetric.

The sheet dimensions increase with increasing electric field E_z, as can be seen from a comparison of the curves in Fig. 16 which were obtained for the same gradient, h_1 = 640 Oe/cm. A doubling of the electric field increases the longitudinal dimension $2x_0$ of the sheet by a factor of ~1.4. On the other hand, the sheet dimensions decrease with increasing gradient h of the quadrupole field. Table 1 shows the dimensions of the current region for various magnetic field gradients and for various electric fields.

We conclude from the circumstance that the current region is not cylindrically symmetric that the concentration of the current near the null line cannot be explained simply on the basis of a decrease in the plasma conductivity in a transverse magnetic field [83], as has been observed in several experiments [84, 85]; this localization is apparently due instead to hydrodynamic plasma flow near the null line. Let us compare the small dimension $2y_0$ of the current sheet found experimentally with the radius of that region near the null line within which an isolated ion of mass M_i does not drift in the (x, y) plane in the crossed electric and quadrupole magnetic fields but instead moves along the electric field E_z, thereby producing a current along the null line. The radius of this region can be found by calculating the trajectories of particles in a magnetic field with a null line, with an electric field imposed along the null line [86], on the basis of the exact equations; the result is

$$r_c = \left(\frac{2M_i c^2}{e} \frac{E}{h^2}\right)^{1/3}. \qquad (67)$$

The values of r_c are shown in Table 1 for the experimental values of h and E; these values were calculated for H^+ and C^+ ions. Comparison of the values of y_0 and r_c in Table 1 shows that the sheet thickness $2y_0$ is less than $2r_c$, even for the atomic hydrogen ion, H^+. Apparently, because of the small Debye length r_d and the appearance of the polarization electric field, the plasma ions continue the two-dimensional motion in the (x, y) plane following the electrons and thus cannot produce a current along the null line* (the Oz axis). But in this situation the plasma density outside the current region, $|y| > y_0$, would be much smaller than that in the sheet; i.e., the original density profile (Figs. 8 and 9) would be deformed by the two-dimensional plasma motion in the (x, y) plane. We previously [87] found qualitative evidence that all the plasma collects in the current region.

Let us examine the magnetic field configuration near the current sheet. Figures 15 and 16 show lines $H = h_1 r$ of the initial quadrupole field with a gradient $h_1 = 640$ Oe/cm. The magnetic field which actually exists near the sheet is the sum of the quadrupole field and the field produced by the plasma current, which is measured by the magnetic probe; the H_x components of these fields are parallel, but their H_y components are antiparallel. Thus the current increases H_x and reduces H_y, so the magnetic lines of force tend to become stretched out along the current sheet. However, as we see by comparing the magnetic field of the plasma current with the field $H = h_1 r$, the field produced by the plasma current is much weaker than the quadrupole field everywhere except right at the null line. Accordingly, the plasma current causes only a very slight change in the initial configuration of the magnetic field under these experimental conditions; the lines of force intersect the current region, and the magnetic field normal to the sheet is nonvanishing, so that the resulting sheet is not a neutral current sheet.

The size of the current region, determined from the magnetic measurements, can be used to calculate the average plasma conductivity over the cross section of the sheet, $\bar{\sigma}$; the results are shown in Table 1: $\bar{\sigma} = (2.5\text{-}4) \cdot 10^{12}$ esu, or smaller by a factor of ~ 50 than the conductivity of the original plasma. Within the errors, the average conductivity over the sheet agrees with the conductivity of the null line, σ_0, obtained from measurements of the magnetic field of the plasma current and calculation of the current density j_{z0} at the null line:

$$\sigma_0 = j_{z_0}/E_z, \tag{68}$$

$$j_{z_0} = \frac{c}{4\pi}(\operatorname{rot} H)_0 = \frac{c}{4\pi}\left(\frac{\partial H_{y'}}{\partial x'} - \frac{\partial H_{x'}}{\partial y'}\right). \tag{69}$$

The values of σ_0 for various experimental conditions are also shown in Table 1: $\sigma_0 = (2.5\text{-}4) \cdot 10^{12}$ esu. From these results we can conclude that the conductivity of the plasma in the current sheet is lower than the Coulomb conductivity and is governed by instabilities driven by the current flowing in the plasma. The effective collision frequency, which is related to the scattering of electrons by the microscopic fields of the waves excited in the plasma,

$$\nu_{\text{eff}} = \frac{ne^2}{m\sigma}, \tag{40'}$$

is no lower than $\sim 10^9$ sec^{-1} in the experiments. This is the lower estimate of ν_{eff} obtained by substituting the original plasma density at the null line, n_0, into (40'). Actually, the plasma density at the sheet can be higher than n_0, so that we would have $\nu_{\text{eff}} > 10^9$ sec^{-1}.

§ 3. Measurements of the Current Distribution in a Magnetic Field with a Null Line in a Plasma with a Density $n \leqslant 10^{13}$ cm^{-3}; Possible Ways to Produce a Neutral Current Sheet

These measurements showed that, as a result of the two-dimensional plasma flow near a magnetic null line, the plasma current takes the form of a sheet whose width and thickness

* These arguments are valid if the ion collision frequency is small in comparison with v_i/r_c.

differ by a factor of about 2-2.5. However, this sheet is not a neutral current sheet; it represents only a slight change in the original configuration of the magnetic field, with essentially no increase in the magnetic energy density near the null line. This circumstance resulted from the anomalously low plasma conductivity found in these experiments. Because of this conductivity, the frozen-in condition for the magnetic field was not satisfied: The ohmic-dissipation time $t_\sigma \sim 3 \cdot 10^{-8}$ sec was much shorter than the time scales $\tau \sim (1\text{-}2) \cdot 10^{-7}$ sec of the processes under study, and the magnetic field was not carried along with the plasma motion. For the energy density of the current sheet to significantly exceed the energy of the original magnetic field, we would have to satisfy the obvious inequality

$$\sigma E \gg \frac{ch_0}{4\pi}, \tag{70}$$

which, because of the smallness of σ, could not be satisfied even at the maximum values of the electric field produced and even at the minimum values of the magnetic field gradient h (Table 1). This anomalously low plasma conductivity was apparently due to the driving of small-scale instabilities by the current flowing in the plasma. Very probably the ion-acoustic instability was driven in these experiments [49, 50]; this instability should lead to a rapid plasma heating over times $\sim(M_i/M)^{1/2}\omega_{0e}^{-1}$, much shorter than the Coulomb-collision time. The rise time for the electric field, $\sim 10^{-7}$ sec, was large in comparison with the scale times for the wave growth and the plasma heating; i.e., the field was turned on "adiabatically." In this case, however, the plasma temperature should "follow" the changes in the current in the system. Usually the heating mechanism in the case of the ion-acoustic instability automatically maintains the difference between the temperatures of the particle species [59]: $T_e > T_i$. The current velocity of the particles near the null line,

$$u_T = j_{z_0}/ne, \tag{71}$$

did not exceed $\sim 5 \cdot 10^8$ cm sec in these experiments, and the condition for the onset of the ion-acoustic instability, $u_T < v_{Te}$, was apparently also satisfied. To satisfy this condition the electron temperature T_e had to be at least 100 eV; i.e., at least 2% of the absorbed energy had to remain in the plasma [see (66)]. In the growth of ion sound, the current velocity is known to be proportional to the ion-acoustic velocity [49, 50],

$$u_T = \alpha c_s = \alpha \sqrt{T_e/M}, \tag{72}$$

where $a \sim (1\text{-}10)$. Such a relationship between the current velocity and the electron temperature can apparently explain the linear increase in the plasma current with increasing applied voltage (Fig. 14): If the number of particles in the plasma does not change significantly, the temperature should increase in proportion to the increase in the energy fed to the plasma, $T_e \sim U_2^2$; according to (72), we have $u_T \sim (T_e)^{1/2} \sim U_2$, so that the current through the plasma, $\mathcal{J}_2$, should also be proportional to U_2. The plasma temperature in the current sheet can be estimated by equating the ponderomotive forces compressing the sheet along the y axis to the gradient of the gaskinetic pressure in the sheet opposing this compression:

$$F = \frac{1}{c}[\mathbf{j} \times \mathbf{H}] \simeq \frac{1}{4c} j_{z_0} H_x, \tag{73}$$

$$\nabla p \simeq nkT/y_0, \tag{74}$$

$$nkT \simeq \frac{y_0}{4c} j_{z_0} H_x. \tag{75}$$

For the experimental conditions corresponding to the first line in Table 1, for example, we have $y_0 = 0.8$ cm, $j_{z0} = 540$ A/cm^2, and $H_x \simeq H_{x'} + hy_0 = 230 + 510 = 740$ Oe; hence $n(T_e + T_i) \sim$

$5 \cdot 10^{15}$ eV/cm^3. According to this estimate the temperature of the plasma electrons could in fact have been above 100 eV. On the other hand, the turbulent plasma heating by the current and the increase of its gaskinetic pressure would apparently have led to a situation in which the electric current could flow across the magnetic field essentially wherever there was a plasma; i.e., there would have been a restoration of the conductivity [88, 89] in a transverse magnetic field.

As was mentioned in §5 of Chapter 1, the current flow in a transverse magnetic field is accompained by the excitation of wave branches other than the ion-acoustic branch. We did not study the nature of the instabilities driven in the current sheet. The results of importance to us were that the plasma resistance was anomalously high, was of a turbulent nature, and increased with decreasing plasma density and that the poor plasma conductivity apparently prevented the formation of a neutral current sheet.

Under these circumstances we took hope from experiments carried out with a current parallel to a magnetic field [46, 47, 61], which demonstrated that the turbulent conductivity of a plasma increases with increasing plasma density, $\sim\sqrt{n}$. Perhaps the conductivity in our system could be increased by increasing the plasma density, so that the frozen-in condition could be satisfied. It was possible that the conductivity would remain turbulent but increase, and it was also possible that a density increase would lead to an increase in the frequency of Coulomb collisions, so that the instabilities could not occur.

CHAPTER IV

Θ DISCHARGE IN A QUADRUPOLE MAGNETIC FIELD AS A METHOD FOR PRODUCING A DENSE PLASMA

As a result of the experiments described in the preceding chapter, we were confronted with a problem of producing a plasma with a density $n \gtrsim 10^{14}$ cm^{-3} and an electron temperature $T_e \gtrsim 5$ eV in a quadrupole magnetic field. We assumed that under these conditions the plasma conductivity, regardless of its nature, would have to be quite high. Since plasma injectors produce only a limited number of particles, we turned to the breakdown of a neutral gas to produce a dense plasma. Here there was an experimental complication, however: With any reasonable choice of the direction for the ionizing electric field (E_z or E_θ) there was a corresponding component of the quadrupole magnetic field (H_θ, H_r) perpendicular to the electric field and hindering breakdown. In the case of the Θ discharge (with an E_θ ionizing field) the situation is slightly better than with a Z discharge, since in the former case there are regions in which the electric field is parallel to the magnetic field (see Fig. 7b, which shows the lines of force of the quadrupole field), and only near the magnetic "apertures" is the magnetic field strictly perpendicular to the electric field. Furthermore, in a Θ discharge the electric field is maximal at the wall of the vacuum chamber,

$$\tilde{E}_\theta = \frac{r}{2c}\dot{\tilde{H}}_z, \tag{76}$$

so that we would expect breakdown at the wall. Under these circumstances the chamber would presumably be filled with plasma with a density comparatively uniform in the radial direction. In the case of an E_z field, on the other hand, the neutral gas breaks down much more easily near the null line, so that a significant radial drop of the plasma density can occur. These considerations led to our choice of the Θ discharge as a method for producing a comparatively dense plasma in a quadrupole magnetic field.

§ 1. Influence of the Quadrupole Magnetic Field on Gas Breakdown in the Electric Field of a Θ Discharge

The physics of the events leading to the formation of an electron avalanche and the initial gas breakdown under the influence of a fixed or rf electric field has been studied thoroughly (see, e.g., [90-96]). Breakdown occurs if the rate at which electrons are formed by ionization exceeds the rate at which electrons are consumed (the Townsend condition). Then a self-maintained discharge occurs, and the number of electrons increases exponentially with the time:

$$n_e(t) = n_{e0} \exp(n_a \langle \sigma_i v_e \rangle t), \tag{77}$$

where n_{e0} is the number of electrons at time t = 0, $n_e(t)$ is the number at time t, n_a is the neutral gas density, σ_a is the effective cross section for electron-impact ionization of the neutral atom, and v is the electron velocity. The value of $\langle \sigma_i v_e \rangle$ is averaged over the electron velocity distribution. It is assumed here that the basic process leading to the electron avalanche is electron-impact ionization of the neutral atoms. In order to cause ionization an electron must of course have an energy above the ionization potential W_i, but the overwhelming majority of the secondary electrons appear with a far lower energy (Fig. 17), so that an electron avalanche can occur only if these secondary electrons can acquire from the electric field an energy sufficient for subsequent ionization events.

A magnetic field perpendicular to an electric field magnetizes electrons, prevents them from acquiring energy in the electric field, and thus hinders the breakdown of the neutral gas. This is the situation we run into, for example, in the ordinary Θ pinch, in which the oscillating magnetic field H_z of the system prevents neutral gas breakdown in the induction electric field E_θ [97]. Kossyi [98] has shown that the development of ionization processes in Θ disharges is governed primarily by the time intervals in which the magnetic field $\tilde{H}_z$ changes sign and has a very small magnitude – the "nonadiabatic" intervals:

$$0 \leqslant |\tilde{H}_z| \leqslant \sqrt{\frac{2mc}{e}} \sqrt{|\dot{\tilde{H}}_z|}. \tag{78}$$

Only in these time intervals, while the magnetic field is weak, can the transverse electron energy increase rapidly and become sufficient to cause ionization. When the equality opposite (78) holds, on the other hand, the electrons gain a negligible amount of energy, and essentially only those secondary electrons which have an energy W_i from the very beginning can participate in ionization. According to Fig. 17 the relative number of these energetic electrons is extremely small, although it does increase with increasing primary-electron energy. The scale time for the occurrence of breakdown in Θ-pinch discharges is strongly influenced by the initial charged-particle density: If this density is small, and the breakdown does not manage to

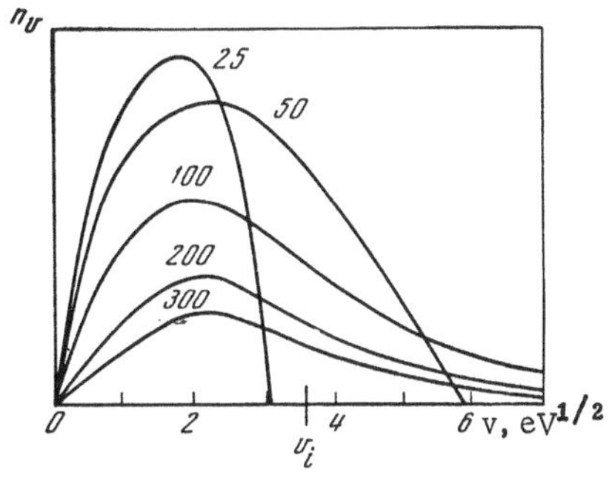

Fig. 17. Velocity distribution of the secondary electrons during electron-impact ionization of hydrogen. The curves are labelled with the energy of the primary electrons, in electron volts [96]. The abscissa shows the velocity of the secondary electrons, while the ordinate shows the number of electrons per unit velocity interval.

occur over the first "nonadiabatic" interval in (78), the breakdown is generally delayed one or several half-cycles with respect to the application of the induction field E_θ [99]. Accordingly, if the ignition of the Θ discharge is to be stable over time there must be a significant preionization ($n \gtrsim 10^{10}$ cm^{-3}).

However, in a strong transverse magnetic field energies above W_i can be acquired in elastic collisions between magnetized electrons and neutral gas molecules. Since the relative change in the transverse electron velocity in such collisions is on the order of the transverse velocity itself, each collision displaces an electron by a distance on the order of the gyroradius, and this displacement can occur along the electric field. The displacements due to repeated collisions with molecules are of a random-walk nature, so that the mean square particle displacement $\overline{r^2}$ over a time t is equal to the sum of the squares of the individual displacements:

$$\overline{r^2} = \nu_{e0} t \left(\frac{v_e}{\omega_H}\right)^2, \tag{79}$$

where v_e/ω_H is the electron gyroradius and $\nu_{e0} = n_0\sigma_{e0}v_e$ is the frequency of elastic collisions between electrons and molecules; in other words, $\nu_{e0}t$ is the number of collisions over a time t.

We introduce a coordinate system whose z axis is parallel to the magnetic field and whose x axis is parallel to the electric field. We assume that the initial particle velocity is low and that the electrons acquire energy as they are displaced along the electric field:

$$W = \frac{mv^2}{2} = eE_x x. \tag{80}$$

If the collision frequency ν_{e0} is independent of the electron energy [this is essentially the case, for example, for hydrogen and helium at energies from ~3 to ~40 eV [95], over which range we have $\nu_{e0} = 5.9 \cdot 10^9 p(H_2)$ and $\nu_{e0} = 2.4 \cdot 10^9 p(He)$ (where p is in torr)] we find from Eqs. (79) and (80)

$$d(x^2) = \frac{1}{2} d(r^2) = \frac{\nu_{e0} v_e^2}{2\omega_H^2} dt,$$

or

$$d(W^2) = \frac{e^2E^2\nu_{e0}}{2\omega_H^2 m} W dt.$$

Then the energy is given as a function of the time by

$$W(t) - W(0) = \nu_{e0} \frac{mu_E^2}{2} t \equiv \frac{mc^2E^2}{2H^2} \nu_{e0} t, \tag{81}$$

where $u_E = cE/H$ is the electric drift velocity of the particles.

As they move along the electric field the particles thus acquire energy more effectively the higher the mean velocity in the direction perpendicular to the electric field (the drift velocity u_E); this is a typical circumstance for diffusion processes in a strong magnetic field [100]. We see from Eq. (81) that, through elastic collisions, an electron in a strong transverse magnetic field acquires the energy required to cause ionization over a time which is shorter the stronger the electric field, the higher the collision frequency (or the higher the neutral gas pressure), and the weaker the magnetic field.

The processes leading to the development of an electron avalanche and gas breakdown in a quadrupole magnetic field under the influence of an electric field E_θ are completely different in nature in regions in which the quadrupole field prevents breakdown (the magnetic "apertures") and in regions in which the electric field E_θ is parallel to the magnetic field. In the latter regions the breakdown occurs as in an ordinary Θ discharge; the nonadiabatic intervals in (78) with respect to the magnetic field of the Θ discharge are important. At the apertures of the quadrupole field, ionization can be caused by the particles which acquire energy in displacement along the electric field in several elastic collisions or by the relatively few secondary electrons which appear with energies above the ionization potential. Then it seems obvious that with an increase in the rate at which an electron avalanche develops in a region with $E_\theta \parallel H_{qu}$ there will be an increase in the rate of breakdown in regions with $E_\theta \perp H_{qu}$. Then for gas breakdown in a quadrupole magnetic field the electric field E_θ must be strong, and there must be a pronounced preionization.

§ 2. Experiments on Gas Breakdown in a Quadrupole Field

In these experiments the vacuum chamber is evacuated to a pressure $p_0 \simeq (2\text{-}3) \cdot 10^{-6}$ torr and then filled with the test gas (H_2, He, Ne, Ar, or air) to a pressure of 10^{-1}-10^{-4} torr by means of a needle valve. The gas is admitted at the same end of the chamber at which the evacuation is carried out, in order to minimize the pressure gradient along the length of the apparatus.

The Θ discharge is excited by 12 turns ~1.5 cm apart wound around the outside of the chamber (Fig. 18). The turns are connected in parallel. A KPM low-inductance capacitor ($C = 0.1\ \mu F$, $U \leq 50$ kV) is discharged into these turns by a controllable low inductance switch. The frequency of the system is $f = 1.55$ MHz, and the electric field produced at the chamber wall is $E_\theta \leq 500$ V/cm.

The preionization is provided by four arc injectors [101], inserted into the vacuum chamber from its end, which inject plasma under the Θ discharges parallel to the quadrupole mag-

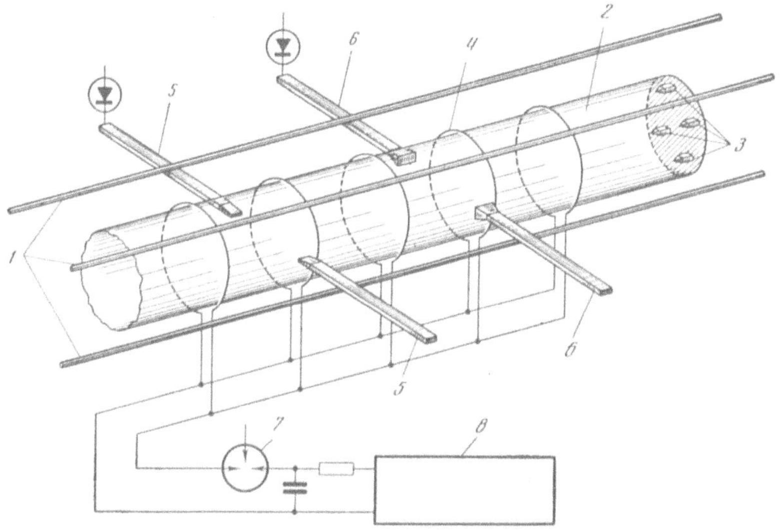

Fig. 18. System for producing the plasma in a quadrupole magnetic field by means of a Θ discharge with pronounced preionization. 1) Quadrupole-field conductors; 2) vacuum chamber; 3) arc injectors; 4) turns of the Θ discharge; 5) microwave antennas, λ = 8 mm; 6) microwave antennas, λ = 2 mm; 7) pressurized switch; 8) 50 kV power supply.

netic field (Fig. 18). A capacitor with C = 0.1 μF, charged to U = 6 kV, is discharged into each of these injectors. Plasma is injected into the discharge gaps of all the injectors from a single ignition device, so that the injectors operate essentially simultaneously (the spread in operating times is less than 10^{-7} sec). The injected plasma moves along the axis of the vacuum chamber; the time delay between the operation of the injectors and the operation of the Θ-discharge switch (~5 μsec) is determined by the time the maximum plasma density arrives in the region in which the Θ discharge is ignited. The arc injectors and the Θ discharge operate at the peak of the essentially steady-state quadrupole magnetic field. The actual occurrence of the gas breakdown and the plasma production are inferred from the appearance of a glow in the chamber and from the cutoff of a microwave signal [102] at a wavelength of λ = 8 mm ($n_{cr} \simeq 1.7 \cdot 10^{13}$ cm^{-3}). When the gas breakdown occurs and the glow appears the plasma density in the quadrupole field is usually above $1.7 \cdot 10^{13}$ cm^{-3}, and the microwave signal is cut off. A standard GZ-30A generator is used to produce the microwave signal at λ = 8 mm. The transmitting and receiving antennas are dielectric (Teflon) rods inserted into metal waveguides. The envelope of the microwave signal is distinguished with a D-402 crystal detector and fed to the amplifier of an S1-29 or S1-42 oscilloscope. The density of the resulting plasma is also measured by a microwave probing signal at λ = 2.2 mm ($n_{cr} \simeq 2 \cdot 10^{14}$ cm^{-3}). This signal is generated by an LOV-2M tube; the waveguide system is made from standard parts of a KPM-1 set and the transmitting and receiving antennas are pyramidal horns. The signals are detected by D-407 crystal detectors, designed for the 4-mm range. Their sensitivity at $\lambda \sim$ 2 mm is far lower than the sensitivity of standard detectors for this range, e.g., the D-1529, but such standard detectors could not be used in our experiments because of their poor noise resistance.

In the absence of a quadrupole magnetic field the gas breakdown caused by the electric field of the Θ discharge occurs even without preionization. In this case the breakdown occurs not during the first half-cycle of the field E_θ, but at a later time. However, at the minimum magnetic field gradient used in these experiments, h_0 = 500 Oe/cm, no breakdown occurred without preionization. If any one of the four injectors was not operating, the breakdown was greatly hindered. Accordingly, this preionization is crucial for igniting a Θ discharge in a quadrupole magnetic field. The density of the plasma injected by the four arc injectors into the region in which the Θ discharge was ignited was n ~ 10^{12} cm^{-3}.

The measurements revealed that for each electric field E_θ, for each type of gas, and for each gas pressure there is a corresponding threshold magnetic field gradient h_{max} which separates the range of magnetic fields in which gas breakdown occurs ($h < h_{max}$) from fields in which breakdown does not occur ($h > h_{max}$) (Fig. 19). We see from the curves in Fig. 19 that gas breakdown in a quadrupole magnetic field requires an electric field Θ which is stronger the higher the magnetic field gradient, the lower the neutral gas pressure, and the smaller the ionization cross section of the atoms of the particular gas (cf. the curves for He and Ar). Accordingly, as E_θ is strengthened, the ranges of the magnetic field and the gas pressure over which the discharge occurs become broader.

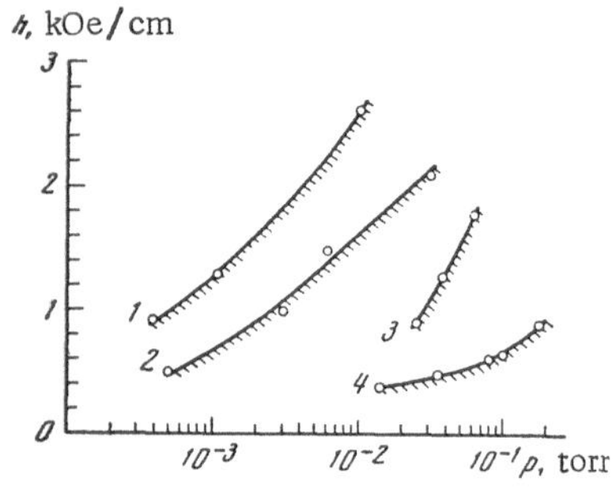

Fig. 19. Gas breakdown by the electric field of a Θ discharge in a quadrupole magnetic field. The curves separate the region of magnetic field in which breakdown occurs ($h < h_{max}$) from the region in which breakdown does not occur ($h > h_{max}$). 1) Ar, E = 430 V/cm; 2) H_2, E = 350 V/cm; 3) He, E = 450 V/cm; 4) H_2, E = 250 V/cm.

Fig. 20. Hydrogen breakdown in a quadrupole magnetic field caused by the electric field of a Θ discharge. The figure gives h_{max} as a function of the electric field of the Θ discharge, with the hydrogen pressure as the parameter. 1) $3 \cdot 10^{-2}$ torr; 2) $6 \cdot 10^{-3}$; 3) $3 \cdot 10^{-3}$; 4) $5 \cdot 10^{-4}$ torr.

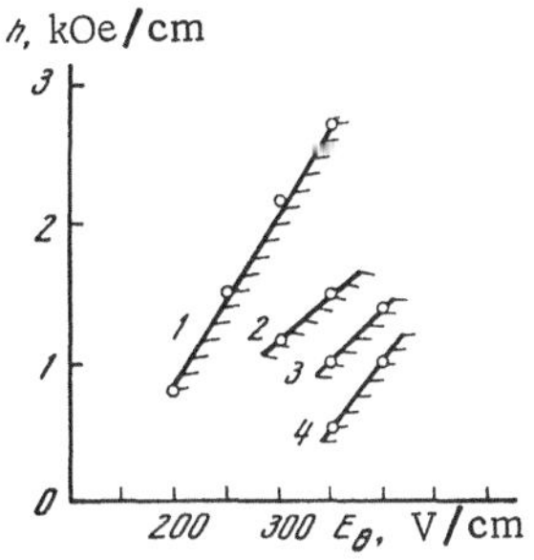

Figure 20 shows this threshold gradient h_{max} as a function of the electric field of the Θ discharge, E_θ, for various pressures of neutral hydrogen. The gradient h_{max} increases essentially linearly with increasing E_θ and also increases with increasing gas pressure. These results show that an important factor in gas breakdown in a quadrupole magnetic field is the acquisition of energy by charged particles in regions of a strong transverse magnetic field, as these particles are displaced along the electric field due to repeated collisions with neutral gas molecules, as discussed in §1. After the gas breakdown, the electric field of the Θ discharge excites a current $\mathcal{J}_2$, in the plasma; the profile of this current over the chamber cross section is measured by magnetic probes. In the absence of the quadrupole field this current is concentrated near the chamber wall, in a region ~1-1.5 cm thick. When the quadrupole field is present, the current in nearly uniform over the entire cross section. It is also interesting to note that the current of the Θ discharge in the plasma decays at essentially the same time as the current in the external circuit igniting the Θ discharge. Accordingly, after the current stops flowing in the external circuit, the plasma produced by the Θ discharge carries no current, and the quadrupole magnetic field completely penetrates this plasma.

In most of the experiments in which gas breakdown occurred the plasma density exceeded $2 \cdot 10^{14}$ cm^{-3}, according to the cutoff of the microwave signal at $\lambda = 2.2$ mm. This cutoff was observed for 40-50 μsec, i.e., over a time vastly longer than the duration of the currents in the Θ-discharge circuit and in the plasma, which decayed over 10-15 μsec. Accordingly, the plasma could be maintained in the quadrupole field for a long time. Furthermore, this prolonged cutoff verified that the cutoff was in fact due to a plasma density above the critical value, $n \gtrsim n_{cr} \simeq 2 \cdot 10^{14}$ cm^{-3}, rather than to, e.g., attenuation of the microwave signal traversing the plasma because of scattering by turbulent fluctuations of a current-carrying plasma, as observed in several experiments [103-104].

The plasma produced by the Θ discharge is displaced away along the axis of the system from the region where it is formed; i.e., if the discharge is not excited over the entire length of the chamber, the Θ-discharge region serves as a sort of injector, like an electrodeless conical source [98, 99]. The plasma velocity along the chamber axis was determined by measuring the time interval between the appearance of the current in the Θ-discharge circuit and the cutoff of the 8-mm microwave signal at a point 30 cm from the Θ-discharge region. The plasma velocity was found to depend on the nature and pressure of the neutral gas and on the strength of the electric field E_θ; under these particular experimental conditions it lay in the range $v_z \simeq 5 \cdot 10^5$-10^7 cm/sec. This velocity was essentially independent of the strength of the quadrupole magnetic field. The plasma could acquire the initial velocity v_{z0}, e.g., through a rapid radial collapse of a current shell [105]. The subsequent motion across the strong inhomogeneous magnetic field could have been made possible by the plasma polarization, as discussed above (§3 of Chapter II; see also [69]).

These results show that a plasma with a density $n > 2 \cdot 10^{14}$ cm^{-3} can be produced in a strong quadrupole magnetic field by means of a Θ discharge and preionization. When the magnetic field gradient is raised above $3.5 \cdot 10^3$ Oe/cm the plasma can apparently still be produced by this method, by increasing the field E_θ and decreasing the frequency of the Θ discharge somewhat.

CHAPTER V

DEVELOPMENT OF A NEUTRAL CURRENT SHEET IN A MAGNETIC FIELD WITH A NULL LINE IN A PLASMA WITH A DENSITY $n \gg 2 \cdot 10^{14}$ cm^{-3}

In the experiments described in this chapter the plasma was produced by a Θ discharge, and its density was generally above $2 \cdot 10^{14}$ cm^{-3}. In a first series of experiments the distance between the electrodes inserted into the vacuum chamber, i.e., the length of the plasma along the magnetic null line, was 44 cm. The system exciting the Θ discharge, 14 cm long, is placed roughly at the center of this interval, and it fills the entire gap between the electrodes with plasma for ~5-20 μsec. The operating time of the induction electric field E_z is chosen to maximize the current in the plasma circuit.

§ 1. Measurement of the Total Plasma Current and the Resistance. Inductance of the Plasma Circuit as a Factor Limiting the Current

Measurements of the plasma current $\mathcal{J}_2$, and of the phase relations between the applied voltage and the current $\mathcal{J}_2$ showed that an increase in the plasma density led to a significant decrease in the resistance R_2 of the circuit with the plasma. Figure 21 shows oscillograms of the currents in the primary circuit of the induction electric field, $\mathcal{J}_1$, and in the secondary circuit, with the plasma, $\mathcal{J}_2$, obtained by Rogowski loops. The phase shift ψ between the currents $\mathcal{J}_1$ and $\mathcal{J}_2$ is approximately π ; i.e., as follows from §4 of Chapter II, the resistance of the secondary circuit is small in comparison with its inductive reactance: $R_2 \ll \omega L_2$. Now the peak value of the current in the plasma circuit is governed by the inductance L_2 of this circuit

$$\mathcal{J}_2 = -\mathcal{J}_1 \frac{\omega^2 M L_2}{(\omega L_2)^2 + R_2^2}\left(1 + i\frac{R_2}{\omega L_2}\right), \qquad (60')$$

$$\mathcal{J}_{20} \simeq \mathcal{J}_{10} M / L_2, \qquad (64')$$

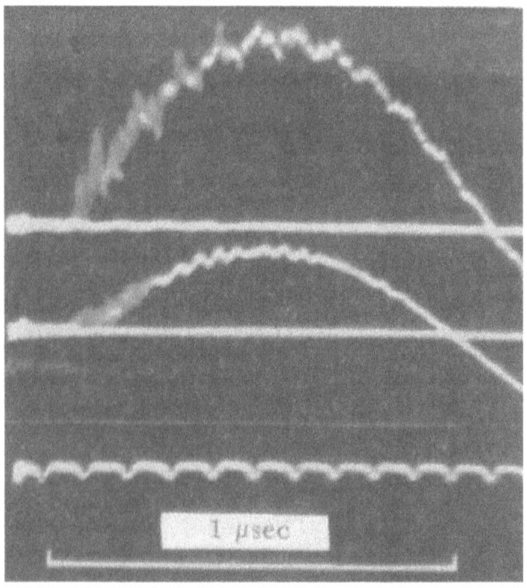

Fig. 21. Oscillograms of the current in the primary circuit for the induction electric field, $\mathcal{J}_1$, and of the current in the plasma circuit $\mathcal{J}_2$. The plasma, $n \gtrsim 2 \cdot 10^{14}$ cm^{-3}, is produced by a Θ discharge in helium at an initial pressure $p_0 = 6 \cdot 10^{-2}$ torr. The gradient of the quadrupole field is $h_0 = 10^3$ Oe/cm.

and the resistance causes the phase shift to differ slightly from π:

$$\psi = \pi - \alpha, \qquad \tan\alpha = R_2/\omega L_2. \tag{82}$$

By measuring $\alpha = \pi - \psi$ we can determined the resistance of the plasma circuit, averaged over the first current half-cycle, and determine the average plasma conductivity $\bar{\sigma}$. Specifically, if we know M and have determined the ratio of peak currents $\mathcal{J}_{10}/\mathcal{J}_{20}$, we can use Eq. (64') to find L_2; after measuring the phase angle α we can use Eq. (82) to determine R_2. The results found from measurements of the peak plasma current $\mathcal{J}_{20}$ during the first half-cycle and the plasma resistance R_2, averaged over the first half-cycle ($T/2 \simeq 1$ μsec), are shown in Table 2 for various experimental conditions. All these results were obtained with the same voltage in the primary circuit: $U_1 = 30$ kV. We note that the resistance R_2 is actually the plasma resistance, since the replacement of the plasma by metal rods resulted in a phase shift $\psi = \pi$ between the corresponding points, within the experimental error. Table 2 shows that R_2 increases with increasing magnetic field gradient h and with decreasing neutral gas pressure p, varying from ~0.03 to ~0.5 Ω. The variation of R_2 with the nature of the gas and its pressure implies that the change in the plasma resistance is caused by a change in the properties of the plasma produced by the Θ discharge. However, this question requires further study. On the other hand, the plasma current $\mathcal{J}_2$ which is limited primarily by the inductive reactance of the circuit, is vastly less sensitive to changes in h and p, varying from 18 to ~12 kA, as we see from Table 2. As the voltage U_1 is increased, i.e., as the electric field E_z is increased, the plasma current increases linearly (Fig. 22).

The voltage U_2 produced in the plasma circuit upon a change in the current $\mathcal{J}_1$ in the primary circuit depends, according to Eq. (65), on the ratio of the mutual induction coefficient M of the circuits and the effective inductance L_{eff} of the primary circuit: $U_2 = U_1 M/L_{eff}$. However, we cannot assume that the entire voltage of the secondary circuit is applied to the plasma, since the voltage distribution depends on the ratio of the inductances of the various circuit elements (the electrodes, the plasma, etc.). The actual voltage across the plasma was determined in special experiments in which the plasma was replaced by metal rods of various shapes; the current ratio $\mathcal{J}_1/\mathcal{J}_2$ was determined and then used in (64') to determine L_2. Comparing the inductance L_2 of the plasma circuit with the inductances L_2 found in these model experiments, we can calculate the fraction of the total inductance of the secondary circuit represented by the plasma inductance L_{pl}. Then we have $U_{pl} = U_2 L_{pl}/L_2$, and the electric field is

TABLE 2

Gas and pressure	h, kOe/cm					
	0.6	0.9	1.3	1.8	2.4	2.85
H_2 $p=10^{-2}$ torr	18 kA < 0.05 Ω	17.5 kA 0.1 Ω	16.4kA 0.2 Ω	14.3 kA 0.35 Ω	—	—
H_2 $p=4\cdot10^{-2}$ torr	—	18 kA 0.05 Ω	—	—	—	—
He $p=4\cdot10^{-2}$ torr	18 kA 0.03 Ω	— —	16 kA 0.10 Ω	— —	— —	— —
Ar $p=4\cdot10^{-4}$ torr	17.5 kA 0.06 Ω	15 kA 0.2 Ω	—	—	—	—
Ar $p=2\cdot10^{-3}$ torr	17.2 kA 0.06 Ω	16.5 kA 0.11 Ω	13.3 kA 0.34 Ω	11.5 kA 0.46 Ω	—	—
Ar $p=1.4\cdot10^{-2}$ torr	17.7 kA 0.03 Ω	—	17 kΩ 0.12 Ω	—	14.5 kA 0.30 Ω	13.2 kA 0.46 Ω
Ar $p=4\cdot10^{-2}$ torr	—	—	—	—	—	16 kA 0.23 Ω

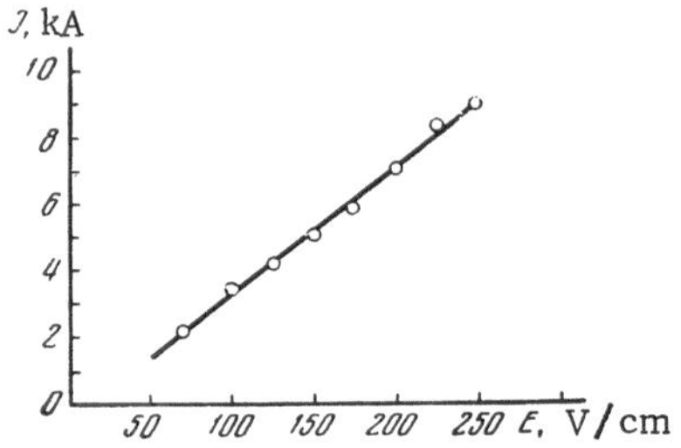

Fig. 22. Peak plasma current $\mathcal{J}_2$ (t = 0.5 μsec) as a function of the induction electric field E_z. The gradient of the quadrupole field is $h_0 = 10^3$ Oe/cm. These experiments involve a Θ discharge in helium at $p_0 = 6 \cdot 10^{-2}$ torr with $n > 2 \cdot 10^{14}$ cm^{-3}.

$E(t) = U_{pl}(t)/l$, where l is the electrode separation. The values of E shown in Fig. 22 and further on were calculated in this manner.

On the basis of these model experiments we can also evaluate the plasma conductivity $\bar{\sigma}$, averaged over space (over the cross section of the current region) and time (over the first current half-cycle). For this purpose we used the metal model for which $(\mathcal{J}_1/\mathcal{J}_2)_{mod} \simeq (\mathcal{J}_1/\mathcal{J}_2)_{pl}$. If we assume the cross-sectional area S of the model to be roughly equal to the cross-sectional area of the current-carrying region in the plasma, and if we determine R_2 from the phase shift between currents $\mathcal{J}_1$ and $\mathcal{J}_2$ [see (82) and Table 2], we find

$$\bar{\sigma} = 9 \cdot 10^{11} \frac{l\,(\text{cm})}{R_2\,(\Omega)\; S\,(\text{cm}^2)} \tag{83}$$

Estimates of this type show that for essentially all values of R_2 we have $\bar{\sigma} \geq 10^{14}$ esu. Although the current distribution in the metal must be quite different from that in a plasma in a quadrupole magnetic field, this estimate does give us the order of magnitude of the conductivity $\bar{\sigma}$ and does show that an increase in the plasma density in fact leads to a significant increase in its conductivity. A more accurate value of $\bar{\sigma}$ can be found by determining the actual dimensions of the current region, e.g., with magnetic probes.

§ 2. Properties of a Fast Magnetosonic Wave Converging on a Magnetic Null Line [106]

We used shielded magnetic probes to determine the spatial profile of the magnetic field of the plasma current (see Chapter III, §2). These probes could be moved along two mutually perpendicular lines in the (x, y) plane, intersecting the null line (the Oz axis) at small angles, ~10°, with respect to the x and y axes (see the diagram in Fig. 24). As in the experiments described in Chapter III, we introduce a coordinate system x', y', z rotated ~10° around the z axis with respect to the x, y, z system, so that the lines along which the magnetic probes are moved coincide with the Ox' and Oy' axes. The output signals from these magnetic probes, integrated by RC circuits with a time constant $\tau \simeq 5 \cdot 10^{-6}$ sec, are fed to the input of a UZ-5 wide-band amplifier and then to one set of plates of an OK-21 dual-trace pulse oscilloscope. A reference signal – the current $\mathcal{J}_1$ or the plasma current $\mathcal{J}_2$ – is fed to the other set of plates. The actual magnetic field is the sum of the original magnetic field with the null line and the magnetic field produced by the plasma current, measured by the magnetic probes.

Figure 23 shows the time dependence of the component $H_{x'}$ of the magnetic field of the plasma current for various positions of the probe along the y' axis. Comparison shows that the signals obtained from probes at different points are not in phase. There is a clearly defined time separation between the appearance of the signal corresponding to the magnetic field of the plasma current at the plasma boundary (y' = −29 mm) and at points far from the boundary; the closer the corresponding points to the null line (y' = 0), the later the magnetic field associated with the plasma current appears at this point. These oscillograms thus demonstrate the propagation of a wave, which is converging from the plasma boundary on the magnetic null

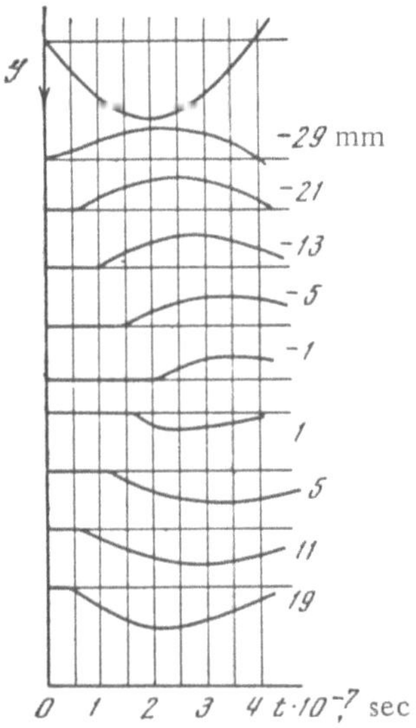

Fig. 23. Oscillograms of the component H_x of the magnetic field of the plasma current for various positions of the magnetic probe along the Oy' axis (the positions are the curve labels). $h_0 = 10^3$ Oe/cm; ⊕ discharge in air; $p_0 = 5 \cdot 10^{-2}$ torr; E_z = 200 V/cm.

line. The time delay between the instants at which the magnetic-probe signal appears at two points is evidently a measure of the propagation velocity of the wave. Measurements carried out with various gradients of the original magnetic field showed that the wave velocity increases roughly in proportion to this gradient.

Figures 24-26 show the spatial profile of the magnetic field of the plasma current for various gradients and at various times after the electric field E_z (parallel to the null line)

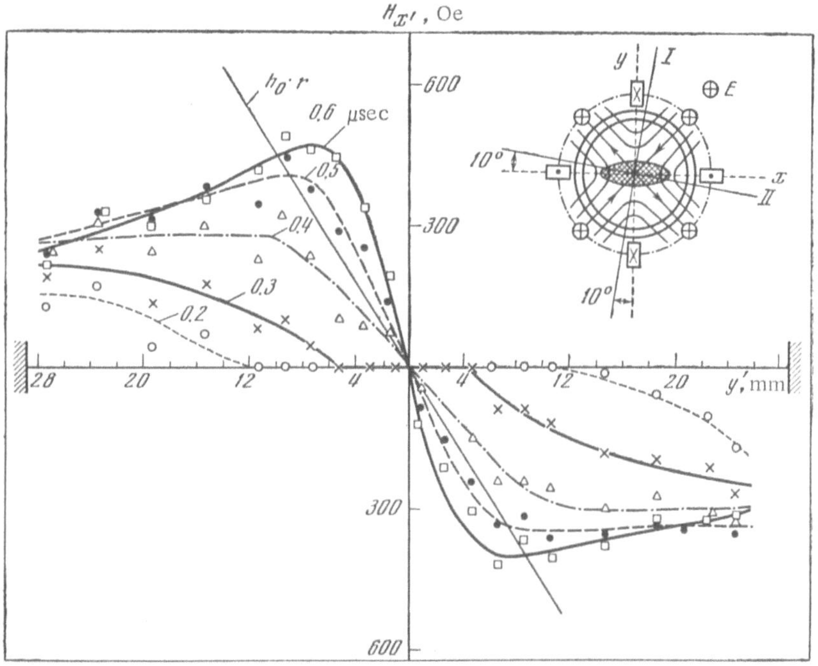

Fig. 24. Profile of the component $H_{x'}$ of the magnetic field of the plasma current along the y' direction at various times after the electric field E_z is applied. The gradient of the quadrupole field is h_0 = 440 Oe/cm; the plasma is produced by a ⊕ discharge in helium; $p_0 = 10^{-1}$ torr; the plasma density is $n > 2 \cdot 10^{14}$ cm^{-3}; E_z = 140 V/cm.

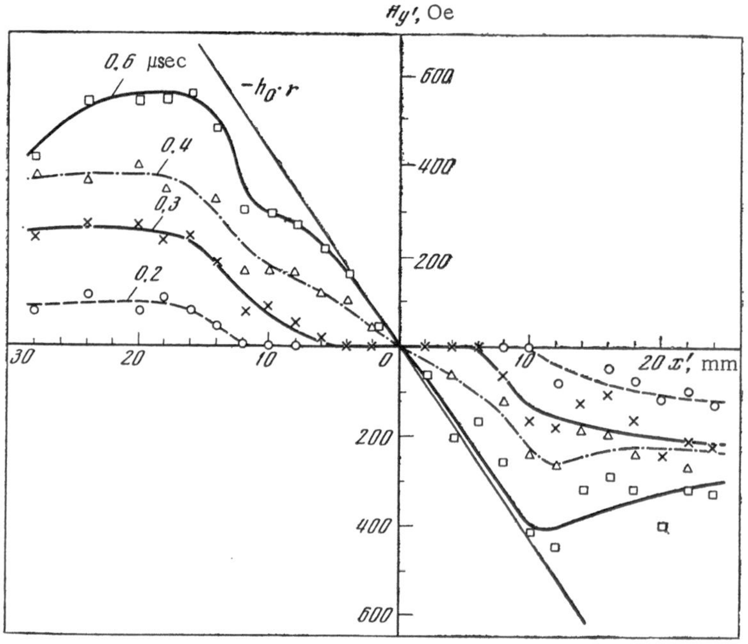

Fig. 25. Profile of the component $H_{y'}$ of the magnetic field of the plasma current along the x' direction at various times after the electric field E_z is applied. The experimental conditions are the same as for Fig. 24.

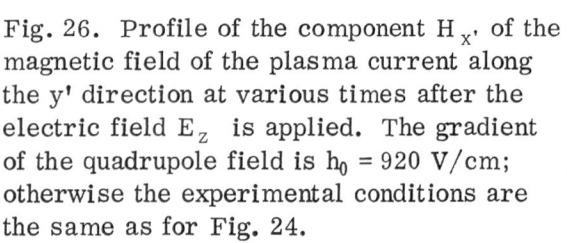

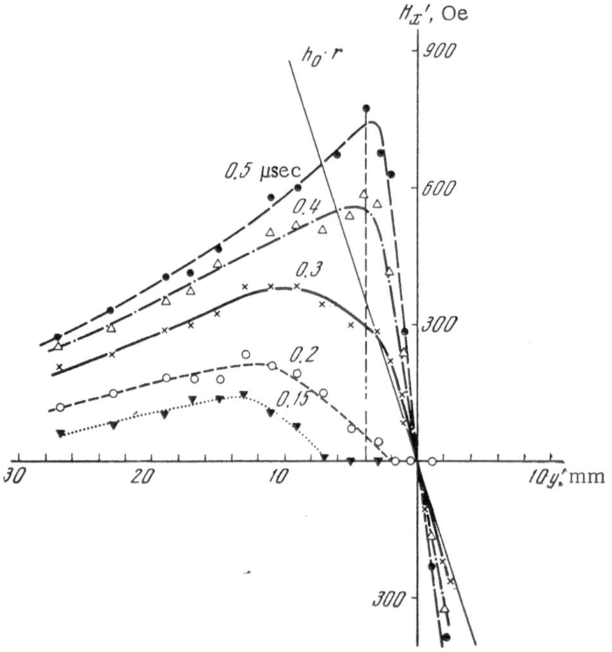

Fig. 26. Profile of the component $H_{x'}$ of the magnetic field of the plasma current along the y' direction at various times after the electric field E_z is applied. The gradient of the quadrupole field is h_0 = 920 V/cm; otherwise the experimental conditions are the same as for Fig. 24.

is applied. These figures show the profiles of the magnetic field components perpendicular to the line along which the magnetic probe was moved; the probe was moved along the radius of the chamber, and the H_φ component of the magnetic field was recorded [either $H_{x'}(y')$ or $H_{y'}(x')$]. The magnetic measurements were very reproducible, so that the profile of the plasma current was recorded successively at various positions of the magnetic probe. Each point on the curves shown in Figs. 24-26 was obtained by averaging from three to five oscillograms. Figure 24 shows that in a magnetic field with an initial gradient $h_0 = 440$ Oe/cm the wave arrives at the null line ~0.4 μsec after E_z is applied; different positions of the front of the propagating wave correspond to the two preceding times shown in Fig. 24 (t = 0.2 and 0.3 μsec). Measurements carried out in two mutually perpendicular directions (Figs. 24 and 25) showed that before the wave reaches the null line (t = 0.2 or 0.3 sec) the magnetic field profile is symmetric with respect to rotation through $\pi/2$; both the magnetic field and the wave velocity are essentially the same. As the gradient of the magnetic field is increased, the wave velocity increases (cf. Fig. 24, with $h_0 = 440$ Oe/cm, and Fig. 26, with $h_0 = 920$ Oe/cm). We see from Fig. 26 that the wave has reached the null line by t = 0.25 μsec. Moreover, the wave velocity at the plasma boundaries, i.e., where the magnetic field is strong, is higher than near the null line. In other words, the wave velocity at each point is proportional to the local magnetic field.

In the stage in which the wave propagates from the plasma boundary to its center, the magnetic field in the wave increases as the null line is approached. However, this field is small in comparison with the original magnetic field $H_0 = h_0 r$, as is clearly seen in Figs. 24-26.

A characteristic feature of the wave which was detected is the absence of a radial component of the magnetic field: $H_r = 0$. The measurements showed that a radial component of the magnetic field of the plasma current appears only after the wave reaches the null line. Figure 27 shows oscillograms of two field components, $H_\varphi(t)$ and $H_r(t)$, obtained at the same point, 15 mm from the null line, i.e., at roughly half the radius of the vacuum chamber. Also shown in this figure is an oscillogram of a reference signal — the current $\mathcal{J}_1(t)$. The H_φ component of the magnetic field appeared ~0.1 μsec after the appearance of $\mathcal{J}_1$; this delay corresponded to the arrival of the front of a radially converging wave at the point r = 15 mm. At this time the H_r component was still zero; it appeared only ~0.27 μsec after the appearance of $\mathcal{J}_1$, i.e., ~0.17 μsec after the appearance of the H_φ component. We see from Fig. 26 that by the time the H_r component appears the wave has reached the null line. The absence of an H_r component from the wave converging radially to the null line shows that this wave is cylindrically symmetric. Measurements of the radial profile of the H_φ component allowed us to calculate the current density j_z at the wave front; we found 150 A/cm² for $h_0 = 440$ Oe/cm and t = 0.3 μsec and 220 A/cm² for $h_0 = 920$ Oe/cm and t = 0.2 μsec.

These results lead to the conclusion that the observed wave was a fast magnetosonic wave converging radially and symmetrically toward the magnetic null line. The properties of the wave agree well with the linear solution [6, 14] (see also Chapter I). Consequently, by

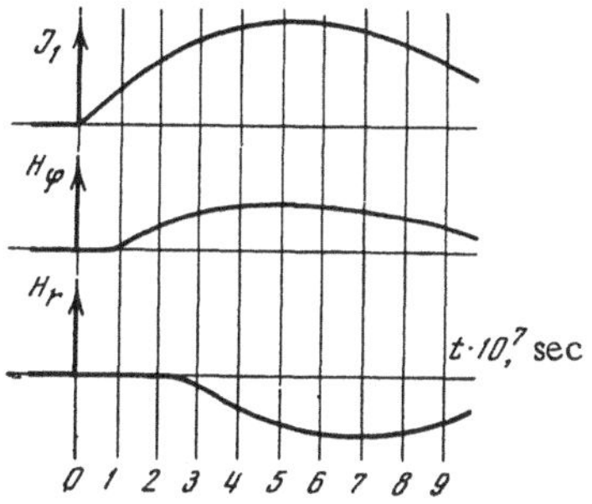

Fig. 27. Oscillograms of the current $\mathcal{J}_1$ in the primary circuit of the induction electric field in the azimuthal H_φ and radial H_r components of the magnetic field of the plasma current at a point r = 15 mm from the null line. $h_0 = 920$ Oe/cm; ⊕ discharge in helium; $p_0 = 10^{-1}$ torr; $n > 2 \cdot 10^{14}$ cm^{-3}; $E_z = 140$ V/cm.

raising the plasma density to a value higher than that in the experiments described in Chapter III we reduced the Alfvén velocity $v_A = H_0/(4\pi\rho_0)^{1/2}$ and found it possible to detect the propagation of a magnetosonic wave, which precedes the appearance and development of the current region near a magnetic null line. However, there is a difference between the magnetosonic waves observed in these experiments and the waves studied in [6, 14] and described by Eqs. (18), (20)-(25) in Chapter I. In a real plasma there are dissipative processes which lead to a damping of waves as they propagate toward the null line. If an electric field which is applied at time $t = 0$ has a value $E_0 = \text{const}$ at the plasma boundary, $r = R$, then in an ideal dissipationless plasma the field is, according to Eq. (18),

$$E(r, t) = -\frac{1}{c}\frac{\partial A_1}{\partial t} = E_0\left(t + \frac{\sqrt{4\pi\rho_0}}{h_0}\ln\frac{r}{R}\right); \quad (84$$

i.e., at each point $r < R$ after the arrival of the magnetosonic wave, i.e., at $t > [(4\pi\rho_0)^{1/2}/h_0] \times \ln R/r$, an electric field of the same strength E_0 as at the plasma boundary should appear. In this case the magnetic field due to the currents in the plasma is, according to (20) and (21),

$$H_\varphi^1 = -\frac{\partial A_1}{\partial r} = \frac{\sqrt{4\pi\rho_0}}{h_0 r} c E_0\left(t + \frac{\sqrt{4\pi\rho_0}}{h_0}\ln\frac{r}{R}\right) \quad (85)$$

and should increase as the wave approaches the null line. At some distance r_1 from the null line the magnetic field in the wave, H_φ^1, can reach the initial level $H_0 = h_0 r$; then the linear solution and thus Eqs. (84) and (85) are no longer valid.

Let us use Eq. (85) to evaluate the quantity

$$r_1 = \frac{\sqrt{cE_0\sqrt{4\pi\rho_0}}}{h_0} \quad (86)$$

for the experimental conditions corresponding to Figs. 24 and 26, with $n = n_{min} \simeq 2 \cdot 10^{13}$ cm^{-3}. With $h_0 = 440$ Oe/cm we find $r_1 = 1.7$ cm, while with $h_0 = 920$ Oe/cm we find $r_1 = 0.8$ cm and $H_\varphi^1(r_1) \gtrsim 700$ Oe. Estimates of this type show that the magnetic field H_φ^1 in a wave without dissipation would be significantly stronger than that in a real wave (Figs. 24 and 26). On the other hand, it is clear from a comparison of H_φ^1 with $H_0 = h_0 r$ (Figs. 24 and 26) that a magnetosonic wave propagating toward the null line is linear essentially all the way to the null line. The perturbation of the vector potential A_1 in the wave and the electric field E_z carried by the wave should also be weaker than the values found from (18) and (84). The field $E_z(r, t)$ can be estimated by assuming that an equation analogous to (85) holds for the propagating wave:

$$E_z(r, t) = \frac{h_0 r}{c\sqrt{4\pi\rho_0}} H_\varphi^1(r, t). \quad (85')$$

With, for example, $t = 0.2$ μsec, $r = 11$ mm, and $h_0 = 920$ Oe/cm (Fig. 26), we find $H_\varphi^1 \simeq 210$ Oe; with $n = n_{min} \simeq 2 \cdot 10^{13}$ cm^{-3} we find $E_z \lessapprox 50$ V/cm, i.e., a field weaker by a factor of ~ 3 than the field at the plasma boundaries, $E_{z0} = 140$ V/cm.

From Eqs. (23) and (24) we find the following equations for the velocity of matter in the magnetosonic wave:

$$v_r = \frac{H_\varphi^1}{\sqrt{4\pi\rho_0}}\cos 2\varphi = \frac{cE}{h_0 r}\cos 2\varphi,$$
$$v_\varphi = \frac{H_\varphi^1}{\sqrt{4\pi\rho_0}}\sin 2\varphi = -\frac{cE}{h_0 r}\sin 2\varphi; \quad (87)$$

i.e., the velocity of matter is equal to the drift velocity in the electric field of the wave, E, and the magnetic field $H_0 = h_0 r$. It also follows from (87) that the ratio of the velocity of matter, $|v| = H^1_\varphi/(4\pi\rho_0)^{1/2}$, to the wave phase velocity $v_A = h_0 r/(4\pi\rho_0)^{1/2}$ is equal to the ratio of the magnetic field in the wave, H^1_φ, to the initial magnetic field, $H_0 = h_0 r$:

$$\frac{|v(r,\ t)|}{v_A} = \frac{H^1_\varphi(r,\ t)}{h_0 r}. \tag{88}$$

This means that in the stage of the experiments in which the magnetosonic wave was propagating, with $H^1_\varphi < h_0 r$ (Figs. 24-26), the velocity of matter in the wave was always lower than the Alfvén velocity. For the example discussed above (Fig. 26: $h_0 = 920$ Oe/cm, r = 11 mm, $h_0 r = 10^3$ Oe, and $H_\varphi = 210$ Oe) we have $|v|/v_A \simeq 0.21$; with $v_{A\max} = 2.4 \cdot 10^7$ cm/sec, corresponding to $n_{\min} = 2 \cdot 10^{13}$ cm^{-3}, we have $|v|_{\max} \simeq 5 \cdot 10^6$ cm/sec.

It also follows from (25) that the original plasma-density profile changes only slightly as the wave propagates. From the measured time required for the wave to propagate between two points we can evaluate the average plasma density between these points. The velocity of the wave converging toward the null line is

$$v_A = \frac{h_0 r}{\sqrt{4\pi\rho_0}} = -\frac{dr}{dt}$$

and the time required for it to propagate between R_1 and R_2 is

$$\Delta t_{1-2} = \frac{\sqrt{4\pi\rho_0}}{h_0} \ln \frac{R_1}{R_2}; \tag{89}$$

hence, knowing h_0, R_1, R_2, M_i, and Δt_{1-2}, we can determine the plasma density. By improving the accuracy of the Δt_{1-2} measurements we can choose the points R_1 and R_2 closer together and thus achieve a more "local" measurement of the density.

§ 3. Development of a Current Sheet near a Magnetic Null Line [107]

After the front of the fast magnetosonic wave reaches the null line, the second stage of the process begins: the formation of a current region near the magnetic null line. The current density at the null line, j_{z0}, increases rapidly, as can be seen from the increased slope of the curves in Fig. 24 for $t \geq 0.4$ μsec and in Fig. 26 for $t \geq 0.3$ μsec in the region in which these curves change sign. Figure 28 shows the change in the current density at the null line of a quadrupole magnetic field with a gradient of 440 Oe cm. This current appears after the arrival of the wave ($t > 0.3$ μsec) and then increases, continuing to increase even after the electric field at the plasma boundary changes direction ($t = 0.6$ and 0.7 μsec). The current distribution in the (x, y) plane is not symmetric with respect to a rotation through $\pi/2$: The

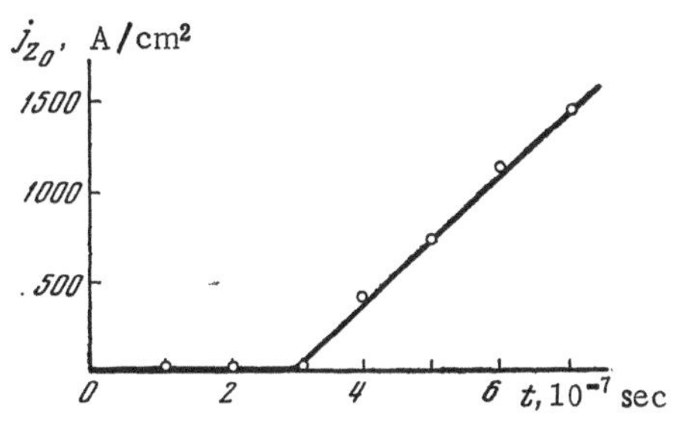

Fig. 28. Time dependence of the current density at the null line $j_{z_0}(t)$, according to the data from the magnetic measurements in Figs. 24 and 25.

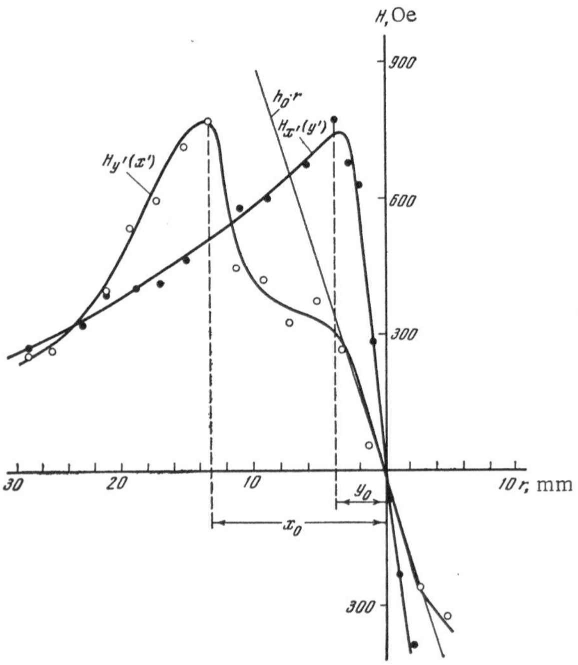

Fig. 29. Measured values of the magnetic field of the plasma current along two mutually perpendicular directions at a time t = 0.5 μsec after the electric field is turned on. h_0 = 920 Oe/cm; ⊛ discharge in helium; $p_0 = 10^{-1}$ torr; $n > 2 \cdot 10^{14}$ cm^{-3}; E_z = 140 V/cm.

current region stretches out along the x axis and contracts along the y axis, i.e., takes the form of a sheet. This result is seen clearly by comparing Fig. 24 with Fig. 25 and the two curves in Fig. 29 obtained as the magnetic probe was moved along and across the sheet. We see from the oscillograms in Fig. 27 that the component H_r appears after some delay with respect to H_φ. This component appears only after the current region begins to form near the null line and demonstrates that this region is not cylindrically symmetric; in particular, it shows that there is a significant component of the magnetic field parallel to the current sheet.

A fundamental difference between this current sheet and that described in Chapter III is a significant distortion of the magnetic field near the current sheet in comparison with the original magnetic field with the null line. We see from Fig. 26 that in a magnetic field with a gradient h_0 = 920 Oe/cm the magnetic field produced at the point corresponding to the thickness of the sheet ($y \simeq 4$ mm) by the current flowing in the sheet is roughly twice the original field at this point; i.e., at the point corresponding to the sheet thickness the magnetic field and its gradient are roughly three times their initial values. Another distinctive feature of the current sheet can be seen from Figs. 25 and 29, which show the spatial profile of the H_y component found by moving the magnetic probe along the current sheet. In the center of the sheet the H_y component of the magnetic field produced by the current is equal in magnitude to the H_y component of the original quadrupole field but opposite in sign; i.e., the resultant magnetic field near the central part of the current sheet does not have a component H_y, normal to the sheet, so that the lines of force in this region stretch out along the sheet without intersecting it. At $r \gtrsim 6$ mm away from the null line an H_y component appears, although it has been greatly weakened by the current flowing in the sheet; i.e., the lines of force begin to intersect the sheet. Comparison of the curves in Fig. 29 also reveals that the dimension of the current sheet along the x axis (x_0 = 14 mm) is greater than that along the y axis (y_0 = 4 mm) by a factor of more than three. The current density at the null line is determined by differentiating the curves for the magnetic field of the current [see (69)]; with h_0 = 920 Oe/cm, E = 140 V/cm, and t = 0.5 μsec (Fig. 29) this current density is $j_{z0} \simeq 4$ kA/cm^2.

However, the magnetic measurements carried out in two mutually perpendicular directions did not give a complete picture of the configuration of the current region; in particular, they did not show whether hydrodynamic flow with slow-mode stationary shock waves occurred during the experiment [10] (the Petschek model; see also §2 Chapter I). The complete pattern of the j_z distribution in the (x, y) plane was determined with the help of a small, movable Rogowski loop, inserted into the vacuum chamber through an electrode at the end placed at a fixed coordinate z, and moved by a positioning device to any desired point in the (x, y) plane. This loop is a toroidal coil of 70 turns, with a major diameter of 5.3 mm and a minor diameter of 1.7 mm. It is first encased in a metal shield with a slit to suppress stray pickup and then in a Teflon shield to prevent electrical contact between the loop and the plasma. The leads are also carefully shielded. The outer diameter of the assembly is 8 mm, while the diameter of its inner aperture is 2.6 mm. The signal from the loop is integrated by an RC circuit with a time constant of 5 μsec; then, like the signal from the magnetic probe, the signal from the loop is fed to the input of a UZ-5 wide-band amplifier and then to the plates of an OK-21 oscilloscope. The reference signal for these measurements is the plasma current $\mathcal{J}_2$, as before. The sensitivity of the loop is ~10 mV/(kA/cm^2). Its frequency characteristic was recorded by means of a special calibration circuit and found to be flat up to ~4 MHz.

Comparison of the geometric dimensions of the Rogowski loop (with an outside diameter of 8 mm) and the dimensions of the current sheet, determined from the magnetic measurements (the thickness was $2y_0 \simeq 8$ mm), shows that the loop could seriously distort the current distribution; thus the results are basically semiquantitative or illustrative. A brief experience with this loop showed that it drew a current from an area larger than the area of its inner aperture. It was found possible to suppress this effect by means of a special thin-walled cap, which serves as a sort of limiter and admits into the inner aperture of the loop only that part of the total current actually corresponding to the area of this aperture. We checked this system by integrating the resulting current-density distribution j_z (x, y) and by comparing with the total plasma current $\mathcal{J}_2$ measured independently. The minimum current density which could be detected by the loop with the particular amplifier and oscilloscope used was 500 A/cm^2, so that the loop did not detect a current in the magnetosonic wave converging on the center of the plasma (see §2 of this chapter); instead it measured the current in the second stage of the process, after the wave had reached the null line and after the current sheet had formed. The measurements of the distribution of the current density in the sheet with the help of this small Rogowski loop confirmed the conclusions reached on the basis of the measurements of the magnetic field of the current.

The signal from the small loop appeared after a delay with respect to the signal from the Rogowski loop measuring the total plasma current J_2; this delay is a measure of the time required for the wave to converge to the null line and marks the beginning of the sheet formation. Figure 30 shows the measured j_z distribution in the (x, y) plane for typical experimental conditions: a quadrupole magnetic field with a gradient of h_0 = 920 Oe/cm, a ⊛ discharge in helium at a pressure of $p_0 = 10^{-1}$ torr, a resulting plasma density above $2 \cdot 10^{14}$ cm^{-3}, and an induction

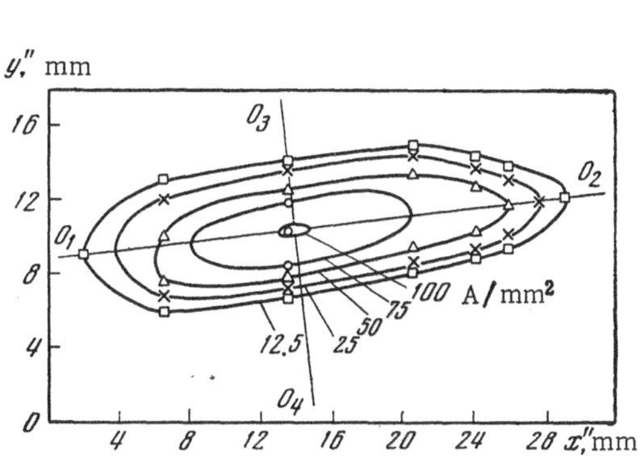

Fig. 30. Profile of the current density in the plane perpendicular to the null line, j_z(x, y) at t = 0.5 μsec. h_0 = 920 Oe/cm; ⊛ discharge in helium; $p_0 = 10^{-1}$ torr; n > $2 \cdot 10^{14}$ cm^{-3}; E_z = 235 V/cm.

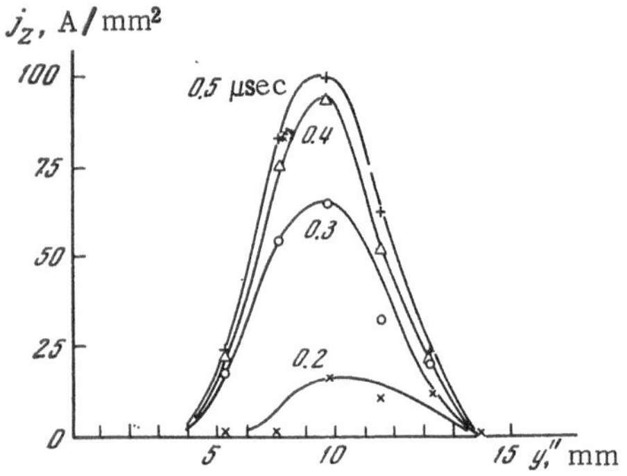

Fig. 31. Current distribution across the sheet, $j_z(y'')$, near the null line at $x'' = 14$ mm, at various times. The experimental conditions are the same as for Fig. 30.

electric field of $E_z = 235$ V/cm. The current distribution shown here corresponds to a time $t = 0.5$ μsec after the field E_z is turned on. The curves in Fig. 30 are lines of equal current density, constructed after measurements of $j_z(y)$ at ten fixed values of x. Figure 30 clearly shows that the current region assumes the form of a sheet; the ratios of the sheet width to the sheet thickness are 3.85, 3.75, 4.1, and 3.8 at the current-density levels 1.25, 2.5, 5.0, and 7.5 kA/cm², respectively. The current density is maximal near the null line (at $x'' = 14$, $y'' = 9$), falling off in all directions away from the null line, most slowly along O_1O_2 and most rapidly along O_3O_4. A very characteristic result is a maximum sheet thickness near the null line (along O_3O_4), decreasing toward the edges of the sheet. This circumstance can also be seen by comparing Figs. 31 ann 32, which show the y'' dependence of the current density at $x'' = 14$ and 24 mm, respectively. These results convincingly show that the Petschek model of slow-mode stationary shock waves [10] is not established in these experiments. Even if we assume that the small Rogowski loop was too crude an instrument to reveal the "fine structure" of the current region, i.e., if we assumed in particular that it could not reveal the "two-humped" nature of the current distribution at a certain distance from the null line, it could still have revealed an expansion of the current region with distance along the line O_1O_2 from the null line. Actually, it revealed the opposite tendency. The clearly defined single maximum in the $j_z(x, y)$ distribution at the null line also refutes the model of [10].

Comparison of the oscillograms obtained with the small loop at the center of the sheet ($x'' = 14$ mm) and at one of the extreme cross sections (e.g., $x'' = 28$ mm) revealed that both the beginning of the signal and the peak at the extreme cross section are delayed 0.1-0.15 μsec with respect to the corresponding events at the center of the sheet; i.e., the sheet expands as time elapses. This behavior is also illustrated by Fig. 33, which shows the profile of the current along the line O_1O_2 at various times. We see that the width (FWHM) of the current distribution increases as time elapses ($\Delta x_{1/2} = 12.5, 16.5$, and 18 mm at $t = 0.2$, 0.3, and 0.5 μsec, respectively). Furthermore, as time elapses the current appears in regions where it was previously lacking; i.e., the base of the $j_z(x)$ curve expands. We also note that as the field E_z increases the width x_0 of the current sheet increases, while the thickness y_0 remains constant (within the experimental errors). This behavior, in particular, furnishes an explanation for the difference

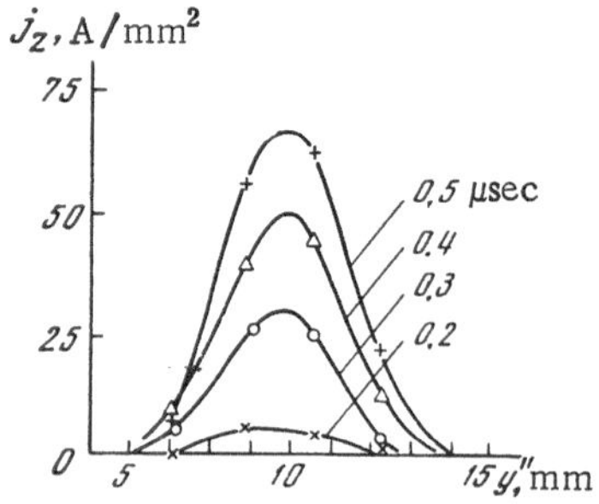

Fig. 32. Current distribution across the sheet, $j_z(y'')$, far from the null line at $x'' = 24$ mm, at various times. The experimental conditions are the same as for Fig. 30.

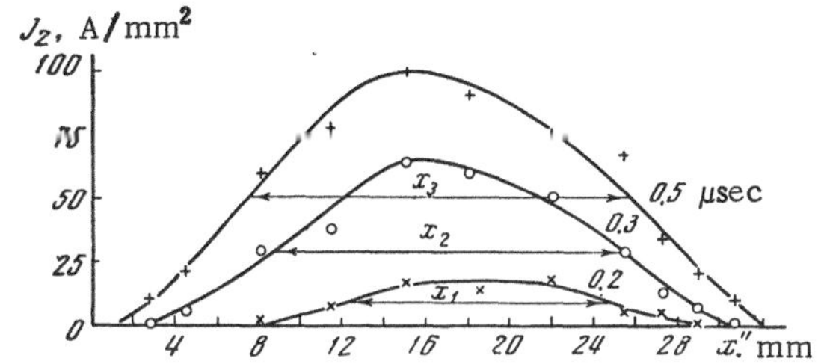

Fig. 33. Current distribution along the sheet (along the line O_1O_2; Fig. 30), $j_z(x'')$, at various times. The half-width of the sheet is: x_1) 12.5 mm; x_2) 16.8 mm; x_3) 18 mm. The experimental conditions are the same as for Fig. 30.

between the values of x_0/y_0 obtained from the magnetic measurements and from the measurements of the current density: The magnetic measurements yield $x_0/y_0 \simeq 3.3$ at $E_z = 140$ V/cm and $t = 0.5$ μsec (Fig. 29), while the current-density measurements yield an average value of $x_0/y_0 = 3.8_0$ at $E_z = 235$ V/cm and $t = 0.5$ μsec (Fig. 30).

We see from Fig. 33 that all the current in this sheet flowed in one direction, i.e., that the return currents which could in principle appear in a neutral sheet [17] did not appear in these experiments.

Measurements of H_y near the sheet (Figs. 25 and 29) show that the sheet cannot be assumed neutral over its entire width. This conclusion also follows from a comparison of the experimental profile of the current density over the width of the sheet, $j_z(x)$ (Fig. 33), with the profile $I_z(x)$ of the surface current density in an infinitesimally thin sheet calculated on the basis of the equations for a neutral sheet (Fig. 6) [17]. The experimental curve of the current density j_z as a function of the coordinate x decays vastly more rapidly away from the null line toward the edges of the sheet than it should in the case of a neutral sheet without return currents (curve A in Fig. 6). In all probability the plasma conductivity was not sufficiently homogeneous over the cross section of the vacuum chamber, with the conductivity at the periphery lower than that at the center. Accordingly, a true netural sheet was still not achieved in these experiments.

§ 4. Production of a Neutral Current Sheet [108]

Now an attempt was made to make the plasma conductivity more homogeneous over the cross section of the vacuum chamber: The distance between the electrodes, inserted from the ends of the chamber into the plasma, was reduced from 44 to 22 cm. This change reduced the time required for the plasma of the Θ discharge to fill the region between the electrodes, so that it was possible to reduce the delay between the end of the Θ discharge and the application of the electric field E_z. Specifically, the electric field in these experimens was turned on essentially immediately ($\tau \lesssim 2$ μsec) after the current stopped flowing in the Θ-discharge circuit. We assumed that a reduction of the delay between the plasma production and the application of the field E_z would be accompanied by a reduction of the plasma loss due to escape to the chamber walls along the magnetic lines of force (Fig. 7); most of this loss was apparently from the peripheral regions.

Measurement of the currents in the primary circuit of the induction electric field, $\mathcal{J}_1$, and in the circuit with the plasma, $\mathcal{J}_2$, and measurements of the phase shifts between the currents showed that the phase angle $\alpha = \pi - \psi$, the difference between π and the phase shift between the two currents, decreased, while the current ratio $\mathcal{J}_2/\mathcal{J}_1$ increased. A decrease in α, according to (82), implies a decrease in the relative importance of the resistance R_2 in the total impedance of the plasma circuit. An increase in $\mathcal{J}_2/\mathcal{J}_1$, according to (64'), implies a decrease in the inductance L_2 and thus an increase in the transverse dimension of the current-carrying region in the plasma (these estimates of course took into account the decrease in the length of the plasma region).

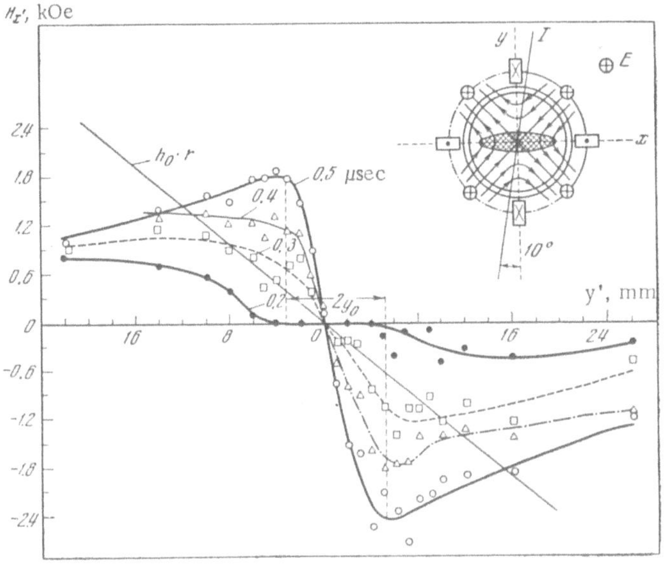

Fig. 34. Profile of the magnetic field of the current sheet, $H_{x'}(y')$, obtained by moving the probe across the sheet. $h_0 = 1.2 \cdot 10^3$ Oe/cm, $\mathcal{J} = 14.3$ kA, $2y_0 = 8$ mm.

Magnetic probes were used for direct measurements of the configuration of the current region under these conditions. Figures 34 and 35 show the results of measurements of the magnetic field of the plasma current obtained by moving the magnetic probe in two mutually perpendicular directions in one typical situation: with a magnetic field gradient of $h_0 = 1.2$ kOe/cm, a Θ discharge in helium at an initial pressure of $p_0 = 6 \cdot 10^{-2}$ torr, a plasma density higher than $2 \cdot 10^{14}$ cm^{-3}, a plasma region 22 cm long, and an induction electric field of E_z = 275 V/cm. Figure 34 shows the profile of the $H_{x'}$ component of the magnetic field of the plasma current along the coordinate y' at various times after E_z is applied. The curve for t = 0.2 μsec corresponds to the wave-propagation stage, while the other curves are successive stages in the evolution of the current sheet. As time elapses the slope of the $H_{x'}(y')$ curves near the null line increases, implying an increase in the current density in this region. At t = 0.5 μsec the sheet thickness $2y_0$ is 8 mm. Comparison of the magnetic field of the plasma current at the boundaries of the sheet, at $y \sim y_0$, with the initial field at this point, $H_0 = h_0 y_0$, reveals that the magnetic field of the current is roughly five times H_0 (Fig. 34), i.e., that the magnetic field and its gradient at the point corresponding to the sheet thickness have increased by a factor of about six. Accordingly, a significant increase in the magnetic energy density was achieved near the null line.

Figure 35 shows the profile of $H_{y'}$ along x', i.e., as the magnetic probe is moved along line II — essentially along the width of the current sheet. Comparing the curves corresponding to t = 0.3 and 0.5 μsec, we see that the sheet width increases as time elapses; at t = 0.5 μsec the measured width is $2x_0$ = 36 mm, i.e., 4.5 times the sheet thickness $2y_0$. The most interesting feature of the resulting profile is that the slope of the $H_{y'}(x')$ curve is equal in magnitude to the slope of the original quadrupole field and has the opposite sign over a distance of ~30 mm; i.e., over this distance the magnetic field near the current sheet does not have a component normal to the sheet, and the lines of force are stretched out along the sheet. Accordingly, a neutral current sheet was produced in this experiment.

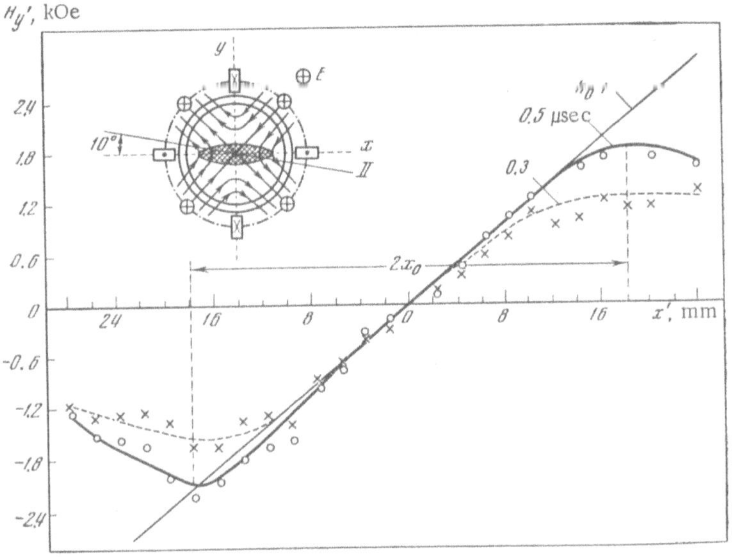

Fig. 35. Profile of the magnetic field of the current sheet, $H_{y'}(x')$, obtained by moving the probe along the width of the sheet at $2x_0 = 36$ mm. The experimental conditions are the same as for Fig. 34.

However, measurements made at even a small angle with respect to the current sheet can understimate the dimension of the sheet along the Ox axis and the distance over which the sheet is neutral, especially since the thickness of the sheet decreases toward its edges, as was shown above. Accordingly, we also measured the H_y component while moving the magnetic probe precisely along the Ox axis; for this purpose the probe was inserted into the vacuum chamber through an aperture in one of the quadrupole-field conductors. These measurements revealed that the half-width of the sheet is $x_0 = 20$ mm and that over the same distance from the null line there is no H_y; i.e., this sheet was neutral over its entire width, and the ratio of its width to its thickness was $x_0/y_0 \simeq 5$. The current density in the sheet near the null line calculated from the magnetic measurements [see (69)] was $j_{x_0} \sim 10$ kA/cm^2 at $t = 0.5$ μsec.

The average conductivity over the cross section of the current sheet was calculated from the measured plasma resistance R_2, as described in § 1 of the present chapter, after the dimensions of the current sheet were determined from the magnetic measurements. Under typical experimental conditions, corresponding to Figs. 34 and 35, the average conductivity over the current region turned out to be $\bar{\sigma} \simeq 2 \cdot 10^{14}$ esu.

Measurements were also made of the profile of H_x along the coordinate x axis $y \to 0$ (Fig. 36). Comparison of the three curves, corresponding to different times, shows that the

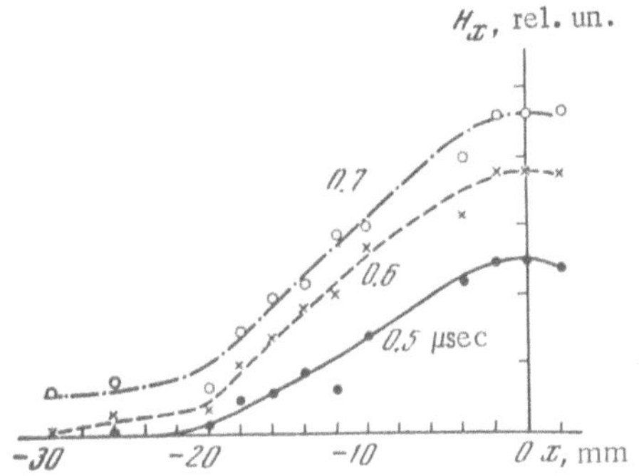

Fig. 36. Magnetic field of the current sheet, $H_x(x)$. The experimental conditions are the same as for Fig. 34.

width of the current sheet increases up to t = 0.7 μsec. This conclusion follows from the appearance of the H_x component at t > 0.5 μsec at those points at which measurements at t = 0.5 μsec had yielded $H_x = 0$ (x = −22 to −30 mm). The significant increase in H at small x was apparently due to a decrease in the thickness of the current sheet over time and to the circumstance that the tube with the magnetic probe passed somewhere within the current sheet (0 < y < 4 mm).

Another important conclusion which can be drawn from the measurements of the longitudinal component of the magnetic field along the sheet is that there are no return currents in the sheet. When we compare the calculated curves for the H_x component of the field of an infinitesimally thin neutral sheet [which are completely identical to the I(x) curves in Fig. 6, because of Eq. (27)], we immediately see that, if a return current appears in the sheet, H_x must change sign at the periphery of the sheet, in the return-current region. However, the experimental $H_x(x)$ profiles reveal no tendency toward such a sign change. Accordingly, no return currents appeared in these experiments under conditions such that a neutral current sheet formed. It may be that the appearance of return currents in the neutral sheet requires additional conditions, e.g., an even higher plasma conductivity.

Experiments carried out with smaller magnetic field gradients h_0 showed that the width of the current sheet increases with decreasing gradient. At $h_0 = 0.57$ kOe/cm, for example, the half-width of the sheet is $x_0 \simeq 28$ mm, i.e., 1.4 times the half-width obtained at $h_0 = 1.2$ kOe/cm (the experimental conditions were otherwise held constant).

As was established in Chapter IV, this Θ discharge system can be used to produce plasma only in a limited range of magnetic field gradients $h < h_{max}$, and for a limited range of neutral gas pressure p_0. For the gases studied, the magnetic field range was widest for argon, as can be seen from, for example, Fig. 19. In order to pursue the study at higher gradients and lower initial pressures we carried out experiments in which the plasma was produced in argon with an initial pressure of $p_0 = 1.4 \cdot 10^{-2}$ torr in a magnetic field with a gradient $h_0 = 2.85 \cdot 10^3$ Oe/cm. An electric field $E_z = 330$ V/cm was applied immediately after the Θ discharge current ended; the total current through the plasma was $\mathcal{J}_z = 13.5$ kA. Figure 37 shows the H_x component of the plasma current as a function of the coordinate y at various times after E_z was applied. The curve for t = 0.3 μsec corresponds to the stage in which the wave propagates from the plasma boundary toward the null line. We see by comparing Fig. 37 with Fig. 34 that a tenfold increase in the mass of the plasma ions led to a significant decrease in the propagation velocity of the magnetosonic wave; this result is particularly noteworthy since the results in Fig. 37 were obtained with a gradient h_0 which was 2.4 times that corresponding to Fig. 34, so that the current sheet began to form the null line only ~0.4 μsec after E_z was applied. By t = 0.7 μsec the current density at the null line reached $j_{z_0} \sim 9$ kA/cm², and the magnetic field

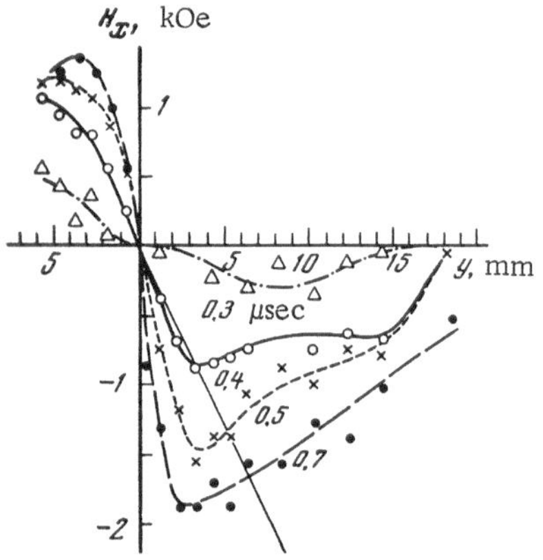

Fig. 37. Profile of the H_x component of the magnetic field of the plasma current along the coordinate y at various times after E_z is applied. The gradient of the magnetic field is $h_0 = 2.85 \cdot 10^3$ Oe/cm; the plasma is produced by a Θ discharge in argon at $p_0 = 1.4 \cdot 10^{-2}$ torr; $E_z = 330$ V/cm.

at the boundaries of the sheet due to the plasma current was $H_{\mathcal{J}} \simeq 1.8 \cdot 10^3$ Oe, or about three times the initial field $H_0 = h_0 r$ at the same point. In other words, the magnetic field at the surface of the sheet was $H_x = H_0 + H_{\mathcal{J}} \approx 2.5 \cdot 10^3$ Oe.

Interestingly, the thickness of the current sheet ($2y_0 = 6$ mm) was slightly less than in the preceding measurements (see Fig. 34, where $2y_0 = 8$ mm); the magnetic field H_x at the boundaries of the sheet was approximately the same in the two cases, $H_x \simeq 2.5 \cdot 10^3$ Oe. Accordingly, a tenfold increase in the mass of the plasma ions did not lead to an increase in the thickness of the current sheet, so that this thickness was independent of the ion gyroradius. We note that the measured sheet thickness, $2y_0 = 6$–8 mm, was several times smaller than the characteristic dimension of the region near the null line within which an isolated ion in the magnetic field of the sheet and in the electric field perpendicular to it should have moved along the electric field, producing a current [see (67)]. This result is analogous to that obtained in Chapter III (Table 1). Apparently because of the relatively high frequency of ion collision, Eq. (67) does not hold in this case, and the ion gyroradius is governed by the ion temperature.

§ 5. Discussion of Experimental Results and Comparison with Theory

The experimental results discussed in the preceding section show that a neutral current sheet can be produced near a magnetic null line in a plasma of sufficiently high conductivity. Let us compare the sheet properties found experimentally with theoretical equations (30)-(32) for an infinitesimally thin neutral sheet in the case in which there are no return currents. For the experimental conditions corresponding to Figs. 34 and 35 the magnetic field gradient is $h_0 = 1.2 \cdot 10^3$ Oe/cm, the total current in the sheet is $\mathcal{J} = 14.3$ kA, the sheet half-width is $x_0 = 20$ mm, the half-thickness is $y_0 = 4$ mm, the ratio of the width to the thickness is $x_0/y_0 = 5$, and the magnetic field at the surface of the sheet is $H_x = H_{\mathcal{J}} + h_0 y_0 = 2.5 \cdot 10^3$ Oe. From Eqs. (30) and (32) we find the sheet half-width b and the field H_x near the sheet to be $b = \sqrt{4\mathcal{J}/ch_0} = 22$ mm and $H_x = 2.67 \cdot 10^3$ Oe ($x \ll b$ in the limit $y \to 0$). Accordingly, the measured magnetic field agrees with the field predicted theoretically, within the experimental error; the experimental sheet width $2x_0$ differs from the theoretical value of 2b by about 10%. This good agreement between experiment and theory, like the direct measurements of the magnetic field, proves convincingly that the experimental sheet was in fact a neutral current sheet.

On the other hand, since Eqs. (30)-(32) give the correct relationship among the experimental parameters, it is sufficient to measure one of the three unknowns H_x, x_0, $\mathcal{J}$ and determine the others from Eqs. (30)-(32). Obviously, the simplest measurement is that of the total plasma current $\mathcal{J}$, which can be carried out in a single shot and which furthermore has all the advantages of contactless methods of measuring plasma properties. Then we would have

$$b = \sqrt{\frac{4\mathcal{J}}{ch_0}}, \qquad H_x^2 = \frac{4\mathcal{J}h_0}{c}. \tag{90}$$

In order to use these relations, of course, we must be sure that a neutral sheet is in fact formed. In this regard the results found in measurements of the magnetic field of the current sheet produced in argon (Fig. 37) are extremely significant. Using Eqs. (90), $h_0 = 2.85 \cdot 10^3$ Oe/cm, and $\mathcal{J} = 13.5$ kA, we find $b = 14$ mm and $H_x = 3.9 \cdot 10^3$ Oe. Thus, the experimental value of the magnetic field near the sheet surface, $H_x \simeq 2.5 \cdot 10^3$ Oe, is less than that calculated from Eq. (90) by a factor of 1.5. The apparent reason for this discrepancy is that in these experiments the condition ensuring that the magnetic field is frozen in the plasma was not satisfied: The conductivity averaged over the cross section of the current sheet is $\bar{\sigma} \sim 5 \cdot 10^{13}$ esu, and the ohmic-dissipation time in (37) for a sheet with a transverse dimension $2y_0 \simeq 6$ mm is $t_\sigma =$

$4\pi\sigma l^2/c^2 \simeq 2.5 \cdot 10^{-7}$ sec, much shorter than the scale time for the sheet formation (Fig. 37). On the other hand, for the neutral sheet of Figs. 34 and 35 we have $\bar{\sigma} \simeq 2 \cdot 10^{14}$ esu and $t_\sigma \simeq 1.8 \cdot 10^{-6}$ sec, or roughly three times the scale times of interest. Comparing these results, we can conclude that the frozen-in condition is a necessary condition for the formation of a neutral current sheet near a magnetic null line.

We did not study the nature of the plasma conductivity, so we cannot say whether the conductivity remained turbulent, as in the experiments described in Chapter III, or whether the increase in the plasma density suppressed instabilities. Further research is required to resolve this question. If we assume that instabilities do not arise and that the conductivity is due to Coulomb collisions, then we can use the value of $\bar{\sigma}$ to determine the temperature of the plasma electrons: $T_e \sim 5\text{-}10$ eV.

The particle density n in the neutral current sheet lies in the range $2 \cdot 10^{14}\ \text{cm}^{-3} < n < 2 \cdot 10^{16}\ \text{cm}^{-3}$. The lower limit $n \simeq 2 \cdot 10^{14}\ \text{cm}^{-3}$ corresponds to the critical density at which the 2-mm microwave signal is cut off (Chapter IV). The upper limit corresponds to 100% ionization of all the neutral gas in the vacuum chamber and to the bunching of all the resulting plasma into a sheet with a cross-sectional area of $S = 4x_0y_0 = 3.2\ \text{cm}^2$.

Finally, on the basis of the measurements of the magnetic field of the neutral current sheet we can estimate the gaskinetic plasma pressure in the sheet: $n(T_e + T_i) \simeq H_x^2/8\pi = 1.6 \cdot 10^{17}\ \text{eV/cm}^3$. Plans call for further experiments to acquire more concrete data regarding the properties of the plasma in the neutral current sheet.

Returning to Eqs. (90), we note that they demonstrate general tendencies in the behavior of the properties of neutral current sheets as the experimental conditions are changed. For example, the sheet width 2b should increase with increasing current $\mathscr{J}$ in the sheet and with decreasing magnetic field gradient h_0, while the magnetic field H_x at the sheet surface should increase with increase in either h_0 or $\mathscr{J}$. Accordingly, if a plasma with a high conductivity is produced in an experiment, and the magnetic field is frozen in this plasma, it is possible to adjust the properties of the neutral sheet in the desired direction; i.e, it is possible to control the sheet properties. This capability leads us on to the next stage of the experiments studying the conditions for the rupture of a neutral current sheet and the generation of pulsed electric fields.

SUMMARY

Theoretical work on the two-dimensional MHD flow of plasma in an inhomogeneous magnetic field with a null line and in an electric field parallel to the null line led to the conclusion that, if the magnetic field is frozen in the plasma, a neutral current sheet should develop near the null line. This sheet would separate regions in which the magnetic fields were equal in magnitude but opposite in direction with lines of force stretched out along the sheet. The formation of a neutral current sheet near a magnetic null line implies a significant increase in magnetic energy density, due to the current flowing in the sheet. The purpose of the experiments described above was to study the conditions for the formation of a neutral current sheet in two-dimensional plasma flow in a magnetic field with a null line and to study the possibility of producing a highly concentrated current and a high magnetic energy density near a null line. In the experimental apparatus used, a magnetic field with a first-order null line and a gradient of up to $5 \cdot 10^3$ Oe/cm near the null line was produced. The electric field parallel to the null line reached 500 V/cm. Experiments were carried out at various initial densities of the plasma produced in the magnetic field with the null line by two methods: with an arc injector ($n \lesssim 10^{13}\ \text{cm}^{-3}$) and by means of a Θ discharge with pronounced preionization ($n \gtrsim 2 \cdot 10^{14}\ \text{cm}^{-3}$). The basic results of these experiments are as follows.

1. The injection method can be used to produce a plasma in an inhomogeneous magnetic field with a null line. Study of the plasma flow in the direction of the null line showed that the minimum transverse dimension of the plasma is ~2-3 cm, the density gradient along the null line is slight, the flow velocity is constant over the cross section, and the plasma motion is due to the appearance of polarization electric fields.

2. With a Θ discharge and pronounced preionization it is possible to produce a plasma with a density $n \gtrsim 2 \cdot 10^{14}$ cm^{-3} in a quadrupole magnetic field. The ranges of the magnetic field gradient and the initial pressure for which gas breakdown and plasma production are possible were determined for various gases.

3. In a plasma with a density $n \lesssim 10^{13}$ cm^{-3} the onset of small-scale instabilities led to the appearance of an anomalous resistance, roughly 50 times the resistance due to Coulomb collisions; as a result, the frozen-in condition was not satisfied in these particular experiments.

Measurements of the magnetic field profile of the plasma current showed that the current was concentrated near the null line, in the form of a sheet with a width-to-thickness ratio of ~2-2.5. The magnetic field of the plasma caused a slight change in the initial magnetic field; i.e., the magnetic energy density near the sheet did not increase. This sheet was not a neutral sheet, however, since there was a magnetic field component normal to the sheet over the entire width of the sheet.

4. It was found that the plasma conductivity could be raised significantly by increasing the plasma density; the conductivity reached $\overline{\sigma} \simeq 2 \cdot 10^{14}$ eus. Now the magnetic field was frozen in the plasma; i.e., the ohmic-dissipation time exceeded the observation time by a factor of three to four.

5. The dynamics of the development of the current region was studied in experiments with a dense plasma ($n \gtrsim 2 \cdot 10^{14}$ cm^{-3}), and it was found that the electric field caused a current and distorted the original magnetic field, initially at the plasma boundaries alone. As time elapsed, this distortion propagated through the plasma as a fast magnetosonic wave converging radially on the null line. The velocity of this wave was equal to the local Alfvén velocity.

6. Under conditions such that the plasma conductivity was $\overline{\sigma} \simeq 2 \cdot 10^{14}$ esu and that the magnetic field was frozen in the plasma, a neutral current sheet forms near the magnetic null line, with lines of force stretched out along the sheet. The width of the sheet is five times its thickness. The basic properties of this neutral sheet – its width, the magnetic field at its surface, and the total current through it – conform well to the equations derived by Syrovatskii [17] for an infinitesimally thin neutral sheet. Direct measurements of the magnetic field of the current sheet, along with the good agreement between experiment and theory, prove that this sheet is in fact a neutral current sheet.

7. Flow with slow-mode stationary shock waves (as in the Petschek model [10]) does not arise near the magnetic null line in these experiments; instead, the neutral current sheet predicted by Syrovatskii [17] develops.

8. A sixfold increase of the magnetic field and its gradient occurs within the neutral current sheet; i.e., there is a significant increase in the magnetic energy density near the null line. The current density at the null line reaches ~10 kA/cm^2.

The author thanks professor M. S. Rabinovich for interest in this study, S. I. Syrovatskii and G. M. Batanov for guidance and useful discussions, and A. Z. Khodzhaev for cooperation and assistance in this study.

Literature Cited

1. G. R. Giovanelli, Nature, 158:81 (1946); Monthly Not. Roy. Astron. Soc., 107:338 (1947); 108:163 (1948).
2. A. B. Severnyi, Astron. Zh., 35:335 (1958); Izv. KrAO, 20:22 (1958).
3. S. I. Syrovatskii, Proceedings of the Fifth All-Union Winter School on Space Physics [in Russian], Inst. Kosmofiziki, Apatity (1968), p. 58.
4. P. A. Sweet, in: Annual Review of Astronomy and Astrophysics, Vol. 7 (1969), p. 149.
5. S. A. Kaplan and V. N. Tsytovich, Plasma Astrophysics [in Russian], Nauka, Moscow (1972).
6. S. I. Syrovatskii, Proceedings of the International Seminar on the Study of Interplanetary Space, Leningrad, 1969 [in Russian], p. 7.
7. J. W. Dungey, Phil. Mag., 44:725 (1953); Cosmic Electrodynamics, Cambridge Univ. Press (1958).
8. P. A. Sweet, Proceedings of the International Astronomical Union Symposium, Vol. 6 (1958), p. 123; Nuovo Cimento, Suppl., 8:188 (1958).
9. E. N. Parker, Astrophys. J., Suppl., 8:177 (1963).
10. H. E. Petschek, Proceedings of the AAS-NASA Symposium on the Physics of Solar Flares, Washington, 1964, p. 425.
11. M. Friedman and S. M. Hamberger, Astrophys. J., 152:677 (1968).
12. M. Friedman and S. M. Hamberger, Solar Phys., 8:104 (1969).
13. S. I. Syrovatskii, Astron. Zh., 43:340 (1966).
14. S. I. Syrovatskii, Zh. Éksp. Teor. Fiz., 50:1133 (1966).
15. V. S. Imshennik and S. I. Syrovatskii, Zh. Éksp. Teor. Fiz., 52:990 (1967).
16. S. I. Syrovatskii, Zh. Éksp. Teor. Fiz., 54:1421 (1968).
17. S. I. Syrovatskii, Zh. Éksp. Teor. Fiz., 60:1727 (1971).
18. B. V. Somov, Dissertation, Moscow Physicotechnical Institute, Moscow (1972).
19. B. V. Somov and S. I. Syrovatskii, Tr. FIAN, 74:14 (1973).
20. S. I. Syrovatskii, Izv. Akad. Nauk SSSR, Ser. Fiz., 31:1303 (1967).
21. V. A. Gribkov, V. M. Korzhavin, O. N. Krokhin, G. V. Sklizkov, N. V. Fillippov, and T. I. Filippova, ZhÉTF Pis. Red., 15:329 (1972).
22. M. Alidieres, R. Aymar, P. Jourdan, F. Koechlin, and A. Samain, Plasma Phys., 10:841 (1968).
23. S. V. Bulanov, S. I. Syrovatskii, A. G. Frank, and A. Z. Khodzhaev, Symposium on Collective Acceleration Methods, JINR, Dubna, 1972 [in Russian], p. 106.
24. A. Bratenahl and M. Yeates, Phys. Fluids, 11:2696 (1970).
25. N. Ohyabu and N. Kawashima, J. Phys. Soc. Jpn., 33:496 (1972).
26. L. D. Landau and E. M. Lifshits (Lifshitz), Electrodynamics of Continuous Media, Addison-Wesley, Reading, Mass. (1960).
27. I. E. Tamm, Basic Theory of Electricity [in Russian], Gostekhizdat, Moscow (1956).
28. E. R. Priest, Monthly Not. Roy. Astron. Soc., 159:389 (1972).
29. S. I. Syrovatskii, Usp. Fiz. Nauk, 62:247 (1957).
30. J. A. Shercliff, A Texbook of Magnetohydrodynamics, Pergamon, New York (1965).
31. S. Chapman and P. C. Kendall, Proc. Roy Soc., A271:435 (1963).
32. S. I. Syrovatskii, Critical Problems of Magnetospheric Physics. Proceedings of the COSPAR Symposium, 1972, IUSTP, Washington (1972), p. 35.
33. H. P. Furth, J. K. Killeen, and M. N. Rosenbluth, Phys. Fluids, 6:459 (1963).
34. G. Laval, R. Pellat, and H. Vuillemin, Proceedings of the 2nd Conference on Plasma Physics and Controlled Nuclear Fusion, Culham, 1965, Vol. II, Vienna (1966), p. 259.
35. V. K. Neil, Phys. Fluids, 5:14 (1962).
36. M. A. Gross and G. Van Hoven, Phys. Rev., A4:2347 (1972).

37. I. M. Podgornyi and R. Z. Sagdeev, Usp. Fiz. Nauk, 98:409 (1969).
38. S. I. Braginskii, in: Reviews of Plasma Physics, Vol. 1, Consultants Bureau, New York (1965).
39. B. B. Kadomtsev, in: Reviews of Plasma Physics, Vol. 4, Consultants Bureau, New York (1966).
40. V. N. Tsytovich, Theory of Turbulent Plasma, Plenum, New York (1974).
41. J. H. Adlam and L. S. Holmes, Nuclear Fusion, 3:62 (1963).
42. S. D. Fanchenko, B. A. Demidov, and D. D. Ryutov, Zh. Éksp. Teor. Fiz., 46:497 (1964).
43. V. A. Suprunenko, E. A. Sukhomlin, and N. I. Reva, At. Énerg., 17:83 (1964).
44. M. V. Babykin, P. P. Gavrin, E. K. Zavoiskii, L. I. Rudakov, and V. A. Skoryupin, Zh. Éksp. Teor. Fiz., 47:1597 (1964).
45. A. M. Stefanovskii, Nuclear Fusion, 5:215 (1965).
46. B. A. Demidov, N. I. Elagin, and S. D. Fanchenko, Dokl. Akad. Nauk SSSR, 174:327 (1967).
47. S. M. Hamberger and M. Friedman, Phys. Rev. Letters, 21:674 (1968).
48. E. K. Zavoiskii and L. I. Rudakov, At. Énerg., 23:417 (1967).
49. L. I. Rudakov and L. V. Korablev, Zh. Éksp. Teor. Fiz., 50:220 (1966).
50. L. M. Kovrizhnykh, Zh. Éksp. Teor. Fiz., 51:1795 (1966).
51. O. Buneman, Phys. Rev. Letters, 1:8 (1958); Phys. Rev., 115:503 (1959).
52. O. Buneman, J. Nucl. Energy, Pt C, 4:111 (1962).
53. V. P. Sizonenko and K. N. Stepanov, Nuclear Fusion, 7:131 (1967).
54. V. I. Aref'ev, I. A. Kogan, and L. I. Rudakov, ZhÉTF Pis. Red., 7:286 (1968); V. I. Aref'ev, Zh. Tekh. Fiz., 39:1973 (1969).
55. J. B. Bernstein, Phys. Rev., 109:10 (1958).
56. A. A. Galeev, D. G. Lominadze, A. D. Pataraya, R. Z. Sagdeev, and K. N. Stepanov, ZhÉTF Pis. Red., 15:417 (1972).
57. D. G. Lominadze, Zh. Éksp. Teor. Fiz., 63:1300 (1972).
58. N. A. Krall and D. L. Book, Phys. Fluids, 12:347 (1969).
59. R. Z. Sagdeev, Proceedings of the 5th European Conference on Controlled Fusion, Grenoble, 1972, Vol. II, p. 105.
60. R. Z. Sagdeev, Proceedings of the 18th Symposium on Applied Mathematics, 1967, p. 281.
61. J. Jancarik and S. M. Hamberger, Proceedings of the 4th European Conference on Controlled Fusion, Rome, 1970, p. 65.
62. S. I. Syrovatskii, in: Solar Flares and Space Research (Z. Svestka and C. de Jager, eds.), North-Holland, Amsterdam (1969), p. 346.
63. S. I. Syrovatskii and A. G. Frank, Preprint No. 120, P. N. Lebedev Physics Institute, Academy of Sciences of the USSR, 1967.
64. V. L. Ginzburg, Propagation of Electromagnetic Waves in Plasma, Addison-Wesley, Reading, Mass. (1964).
65. D. A. Frank-Kamenetskii, Lectures on Plsma Physics [in Russian], Atomizdat., Moscow (1964).
66. L. A. Artsimovich, Controlled Thermonuclear Reactions, Gordon and Breach, New York (1968).
67. T. F. Volkov, Reviews of Plasma Physics, Vol. 4, Consultants Bureau, New York (1966).
68. L. A. Tseitlin and P. L. Kalantarov, Calculation of Inductances [in Russian], Gosénergoizdat, Moscow-Leningrad (1955).
69. S. I. Syrovatskii, A. G. Frank, and A. Z. Khodzhaev, Preprint No. 142, P. N. Lebedev Physics Institute, Academy of Sciences of the USSR, 1972; Zh. Tekh. Fiz., 43:912 (1973).
70. S. Yu. Luk'yanov, I. M. Podgornyi, and V. N. Sumarokov, Zh. Éksp. Teor. Fiz., 40:448 (1961).
71. M. S. Ioffe and R. I. Sobolev, At. Énerg., 17:366 (1964).
72. K. D. Sinel'nikov, N. A. Khizhnyak, et al., Magnetic Traps [in Russian], Naukova Dumka, Kiev (1965), p. 5.

73. B. Logan et al., Phys. Rev. Letters, 28:144 (1972).
74. V. F. Demichev, V. L. Matyukhin, A. V. Nikologorskii, and V. M. Strunnikov, At. Énerg., 19:329 (1965).
75. É. Ya. Gol'ts, V. B. Turundaevskii, and A. Z. Khodzhaev, Preprint No. 129, P. N. Lebedev Physics Institute, Academy of Sciences of the USSR, 1968.
76. G. A. Delone and M. M. Savchenko, Tr. FIAN, 32 (1966); Zh. Tekh. Fiz., 36:1409 (1966).
77. É. Ya. Gol'ts and A. Z. Khodzhaev, Zh. Tekh. Fiz., 39:988 (1969).
78. K. F. Sergeichev, Dissertation, Physics Institute, Academy of Sciences of the USSR, Moscow (1971).
79. B. P. Aseev, Oscillator Circuits [in Russian], Svyaz'izdat, Moscow (1955).
80. S. I. Syrovatskii, A. G. Frank, and A. Z. Khodzhaev, Proceedings of the 4th European Conference on Controlled Fusion, Rome, 1970, p. 66.
81. R. E. Huddlestone and S. L. Leonard (editors), Plasma Diagnostic Techniques, Academic Press, New York (1965).
82. I. M. Podgornyi, Lectures on Plasma Diagnostics [in Russian], Atomizdat, Moscow (1968).
83. S. I. Braginskii and G. I. Budker, in: Plasma Physics and the Problem of Controlled Thermonuclear Reactions (ed. M. A. Leontovich), Vol. I, Pergamon, New York (1958).
84. N. A. Bobyrev and O. I. Fedyanin, Zh. Tekh. Fiz., 31:1309 (1961).
85. M. D. Raizer, P. S. Strelkov, and A. G. Frank, Zh. Tekh. Fiz., 34:1040 (1964).
86. S. V. Bulanov, private communication.
87. S. I. Syrovatskii, A. G. Frank, and A. Z. Khodzhaev, Proceedings of the All-Union Conference on Accelerators, Moscow, 1968 [in Russian], Vol. II, izd. VINITI, Moscow (1970), p. 538.
88. L. Spitzer, Physics of Fully Ionized Gases, Interscience, New York (1956).
89. H. Alfvén and K. H. Falthammar, Cosmical Electrodynamics, Oxford Univ. Press, London (1963).
90. N. A. Kaptsov, Electrical Phenomena in Gases and Vacuum [in Russian], Gostekhizdat, Moscow-Leningrad (1947).
91. A. von Engel, Ionized Gases, Oxford Univ. Press, London (1965).
92. E. W. McDaniel, Collision Phenomena in Ionized Gases, Wiley, New York (1964).
93. H. S. W. Massey and E. H. S. Burhop, Electronic and Ionic Impact Phenomena, Oxford (1952).
94. S. C. Brown, Basic Data of Plasma Physics, MIT Press, Cambridge, Mass. (1959).
95. A. D. MacDonald, Microwave Breakdown in Gases, Wiley, New York (1966).
96. V. L. Granovskii, Electric Currents in Gases [in Russian], Vol. 1, Gostekhizdat, Moscow (1952).
97. H. A. Bodin, T. S. Green, G. B. F. Niblett, and N. J. Peacock, Proceedings of the 4th International Conference on Ionization Phenomena in Gases, 1959, Vol. II, Amsterdam (1960).
98. I. A. Kossyi, Dissertation, Physics Institute, Academy of Sciences of the USSR, Moscow (1968).
99. I. A. Kossyi, I. S. Shpigel', and E. V. Dorofeev, Zh. Tekh. Fiz., 36:881 (1966).
100. V. E. Golant, Usp. Fiz. Nauk 79:377 (1963).
101. W. H. Bostick, Phys. Rev., 106:404 (1957).
102. V. E. Golant, Microwave Methods for Studying Plasmas [in Russian], Nauka, Moscow (1968).
103. A. I. Kislyakov, M. M. Larionov, and V. V. Rozhdestvenskii, Zh. Tekh. Fiz., 37:584 (1967).
104. L. E. Sharp and S. M. Hamberger, Proceedings of the 4th European Conference on Controlled Fusion, Rome, 1970, p. 64.
105. A. C. Kolb, Rev. Mod. Phys., 32(4):748 (1960).

106. A. G. Frank, Proceedings of the International Seminar on Particle Acceleration in Space, Leningrad, 1971 [in Russian], NIIYaF MGU (1972), p. 121.
107. S. I. Syrovatskii, A. G. Frank, and A. Z. Khodzhaev, ZhÉTF Pis. Red., 15:138 (1972).
108. S. I. Syrovatskii, A. G. Frank, and A. Z. Khodzhaev, Proceedings of the 5th European Conference on Controlled Fusion, Grenoble, 1972, Vol. 1, p. 150.